Brain Jolt

A Life Renewed After Traumatic Brain Injury

Second Edition, with Homeopathic Appendix

JoAnn M. Jarvis, R.N., D.H.M.

Brain Jolt: A Life Renewed After Traumatic Brain Injury

Second Edition with Homeopathic Appendix

Foreword by Maria T. Bohle, CCH, RSHom (NA)

This is a personal story of my recovery. It does not provide medical advice. A homeopathic appendix is provided for consideration and study and is intended for homeopaths. The author and publisher are not responsible for any adverse effects or consequences resulting from anything described in this book.

Edited by Robert Barrow, B.L.S.

Cover Design and Art by Amy Chamberlain

Published by JoAnn M. Jarvis

ISBN 978-0-6152-0805-3

*

To:

Bob

Janine

Madeline

Michael G.

Margaret Mathews

Anne Barash, M.D.

Richard Cantrell, M.D.

Christopher Piering, D.C.

Acknowledgments

Special thanks to you, Robert Barrow, for your editing skills, suggestions, endless patience, unique humor, and – oh, yes – the commas I
threw out -- in the writing of this book.

Thank you, Amy Chamberlain, for your artistic gifts, hard work and
patience. You are a joy to work with. You made my vision come to life on
the cover of this book.

Thank you to: Rosalie Muehlberger, Donna Earnest and all of my
family – Bob, Madeline and Janine -- for your encouragement when I
decided to write my story.

Thank you, Maria Bohle, for your suggestion and encouragement to
include a homeopathic appendix for this second edition. I am proud to
have your words grace my book in the Foreword.

Thank you, Roger van Zandvoort, for allowing me to use your rubrics from your excellent repertory, **The Complete Repertory**.

Table of Contents

Foreword

The book you are about to read is most extraordinary. You will learn to understand brain injuries through the patient's experience. The trauma and horror of a person's consciousness locked inside a body they can no longer control is the kind of story an Edgar Allen Poe or certainly *The Twilight Zone* writers would have recognized as a nail-biter.

JoAnn Jarvis is a registered nurse, a caregiver, well versed in modern medicine, whose experiences put her into the archetypal hero category. She is truly a person who has been through hell, fought hard, endured trials, triumphed and returned to tell us about it. Her experience is our treasure; from her pain greater knowledge and understanding were born and she has given that gift to us for the betterment of our world.

Yes, this book is extraordinary! **Brain Jolt** was written for the doctors, for the caretakers, for the families and for the homeopaths who find themselves called to assist those who have experienced brain injuries. **Brain Jolt** is a book all medical practitioners should have on their shelves and this is a reference source that will be part of my library. I know I will refer to it frequently, as a guide to remind me what I need to know to successfully work with both new and old head injuries of all kinds.

Note to homeopaths: JoAnn Jarvis is an experienced professional homeopath. I had the pleasure of meeting JoAnn many years ago, after her trauma and ordeal, when she continued on with the study of homeopathic medicine after it helped her so much. Although this book was not written exclusively for homeopathic practitioners, Mrs. Jarvis was kind enough to write an entire homeopathic appendix to help guide us in case-taking and finding rubrics with people who may not be able to articulate their symptoms. **Brain Jolt** by JoAnn Jarvis will sit in my bookshelf right next to Jollyman's **Asthma**, Borland's **Pneumonias**, Forbister's **Homeopathic Insight Into Cancer** and Perko's book, **The Homeopathic Treatment of Influenza.**

Maria T. Bohle, CCH, RSHom(NA)
CEO, The British Institute of Homeopathy USA

Disclaimer

Please understand, I am not advocating for or even suggesting that you follow the same course of treatment that I describe in my book. The treatments I describe may help your condition or they may adversely affect it. There are no guarantees either way. People's bodies are different, as are their injuries and their individual reactions to medicines or other kinds of treatments. Always check with your physician if there is something you would like to try. Make sure it does not conflict with your condition, your current therapies or treatments. In addition, state laws vary on what alternative/complementary practices are permitted from practitioners other than a licensed medical doctor. Always consult your own physician to be certain what is right for you. This IS NOT a self- help book. A homeopathic appendix has been included in this book for study and use by qualified homeopaths.

Introduction

Your brain has never been in more danger than it is right now. How well, how very well, I know this to be true. Have you ever paused to wonder how dramatically, how horribly, your life could change if your remarkable, yet fragile brain ceased to function normally tomorrow? Many people currently travel that frightening road, and many others soon will. The obvious invading monsters are organic brain syndrome (dementia), Alzheimer's disease, stroke from high blood pressure, a burst aneurysm and traumatic brain injuries. But other causes lurk precariously, too.

You can sustain a brain injury easily from a minor car accident, or you might come home from a war with a traumatic brain injury. If you lapse into a coma, medical professionals will diagnose you correctly. However, should you fail to lose consciousness, avoiding that coma, chances are you will not be diagnosed at all.

As a registered nurse since 1984, currently living and working in Central New York State, I've been married for 28 years, and I've raised two children. My life might sound normal so far, but it hasn't always been so. In fact, there were times when it became a nightmare because, you see, I experienced a traumatic brain injury myself, and I want to share my experience of recovery with you. I think it might help bring awareness to an unrecognized and often undiagnosed problem. I know it will help those who struggle in their own recovery from an (M)TBI or TBI. [I will use the term (M)TBI to designate both Mild Traumatic Brain Injury and Traumatic Brain Injury. These terms will be further explained in the next chapter.] I want my knowledge to assist the professional caregivers and families who live with and care for people with (M)TBI. It's my hope that the information imparted here will also educate people to recognize symptoms and to help those who have not yet been diagnosed. At the very least, you will become sensitive to the constant struggles of a person who lives with (M)TBI.

(M)TBI/Concussion

Years ago, long before the (M)TBI designation gained appreciation, the word "concussion" was the popular medical term. A concussion's basic definition relates to a violent external force to the head, causing the brain to hit up against the inside of the skull, resulting in extensive generalized damage to the brain. This violent movement can be caused by a fall, a blow to the head or even from a whiplash motion which causes the head to flip back and forth, and possibly sideways. In fact, car accidents are a major cause of (M)TBI, along with sports injuries and even amusement park rides. Similarly, it may result from a severe shaking, as in *Shaken Baby Syndrome*. It arises also from injuries sustained during wartime by combat troops. Just having something explode close to your head, surprisingly, can cause serious injury even if it did not administer a direct hit.

Profound brain damage, in addition to occurrence without direct force, can also be devastating without any loss of consciousness. This banging of the brain inside the skull results in a coup/contrecoup injury from the acceleration/deceleration forces upon the brain, and results in diffuse axonal shearing and tearing. Essentially, you have a ripped, bruised brain, along with its blood vessels, accompanied by neuronal death and damaged and ripped connections to other brain cells. This ripping and tearing of brain tissue precipitates a cascade of many biochemical changes in the brain. The brain swelling and biochemical changes produce symptoms. The kinds of symptoms exhibited depend upon which parts of the brain are affected. This is considered an axonal injury with damage to the deep structures of the brain, leading to widespread neurological damage.

Damage becomes more severe and apparent over the course of many weeks, even months, after the injury. The effects are not immediately apparent, sometimes not even over the course of the days immediately following the injury. Usually in traumatic situations other injuries are sustained, and immediate attention is paid to more apparent injuries and pain control. However, when the brain swells because of injury, there is little room for expansion inside the skull. More and more areas of the brain become involved as a result of this swelling and increased pressure. I encourage you to read more on the pathophysiology and signs and symptoms of this condition. Two good sources for research are the Department of Health and Human Services Center for Disease Control and Prevention, and the Brain Injury Association. Both sources make available excellent information and facts about concussion and (mild) traumatic brain injury and traumatic brain injury -- (M)TBI. The website for the CDC is www.cdc.gov/ncipc/. The Brain Injury Association of America's website is www.biausa.org. Their phone number is 1-800-444-6443.

What makes an MTBI different from a TBI? Very simply put: If the person who sustained a brain injury did not go into a coma, that is called a mild traumatic brain injury (MTBI). However, if the person sustained a brain injury and went into a coma, that is a TBI, or traumatic brain injury. [There are other criteria that medically differentiate this as well; please research the *Glasgow Coma Scale* for that specific information, as well as *Grades of Concussion*.] The areas of the brain affected and the cognitive resulting symptoms can be and usually are the same for both TBI and MTBI -- so the word "mild" can be very misleading here. The effects of both can stay with an individual for a lifetime. I am not a physician or an expert in this field and new research and studies are ongoing, so I make no claims that this information will not change. Science and medicine continuously evolve, and therefore I can only offer what I have read in the current literature. But I possess the advantage of living through this experience personally and I work as an R.N. with TBI clients. My focus here is not to teach specified signs and symptoms, but to share my own personal experience of living this out, as well as what helped me.

These days we have more car collisions, in addition to our soldiers returning from the Middle East with head injuries. In the past, people with these severe head injuries died, but now they live because of breakthroughs in medicine and surgery. However, there is no medicine that exists to heal an injured brain. What's more is that no one really understands what this is like unless they have lived through it.

I also want to warn the reader that you are only a "jolt" away from an (M)TBI yourself. No one is safe or immune. Think how you would like to be treated if you or someone you love ends up with this condition.

A Different Way To Heal

My story is unique because I went undiagnosed for so long and used healing modalities other than mainstream conventional medicine. I did not have a doctor directing, coordinating, taking credit or accepting responsibility for what I did, either. Of course, this meant that I had to take total responsibility for what happened or did not happen. There was nobody to blame if things went wrong. I also healed at the deepest level possible, although achieving this took many years. As you will discover, I needed to do something, as my own doctor was not able to recognize the symptoms of my concussion/traumatic brain injury. Another personal fight involved my employer's insurance company, which wanted to label me "fraudulent," and for the longest time assailed me with their harassment techniques. There was no neuropsychologist directing me on how to put my memory or cognition back together and there were no conventional medications taken, although ultimately they were offered.

It is important to realize there are no medications that can specifically treat or cure an (M)TBI. When I finally found a doctor who understood I had a TBI, I refused medications for symptom control, including pain medications. My reasoning was this: The medications would serve only to cover up symptoms, and I wanted to use my symptoms to monitor my healing. The medicines were for symptom palliation and did nothing to heal the brain damage. (Whatever happened to merely resting and letting damaged body parts heal the way our bodies are programmed to heal already?) I have multiple drug allergies and did not want to risk even more, and I did not want to mix side effects into the pain and suffering I already had. As well, I come from a strong family background of using natural means to heal.

I received absolutely no help in the form of nursing, not even home health aides to assist me or my family. I was not offered that kind of help. I was not directed to any specific programs, either, such as a structured day program that dealt with TBI. Instead, when I was diagnosed I was sent home to "rest." I am grateful for that advice, too. Rest is so important. But, for a very long time, I was alone in this situation with a damaged brain. I was told it would take years to recover, and no one knew how I would come out of this. My family was stuck with me, too, and I was very confusing, scary and depressing to be around. I knew I had to do everything possible to heal if I were to come out of this.

Now, I do not see conventional medicine as a negative thing as I am a nurse. I believe in a person's freedom of choice – which goes along with the role of nurse as patient advocate. However the patient wishes to be treated, once he or she is fully informed and understands the consequences, is the way things should be done, that is my opinion.

Conventional medicine offers nothing to cure or heal a brain injury. The symptoms of a bruised brain are treated with medications, without any knowledge of what those medications will do to the brain. They treat insomnia with a drug. They treat "brain fatigue" with a drug. They treat the depression with a drug. They treat the emotional lability with a drug. They treat anxiety with a drug. They treat the brain pain with a drug. Dizziness is very common, and treated with yet another drug. There's a drug for the months-long nausea too. But they don't have a drug for your poor memory, and they don't have a drug for walking around in a daze -- your common everyday life after an (M)TBI. Nor have they a drug for your aphasia, or for your very poor cognitive skills. They don't possess a drug to treat your blank mind.

Most importantly, conventional medicine produces nothing to heal a shattered life or spirit. Neither pills, doctors, nor nurses, can heal shattered relationships with family and friends. There is nothing in medicine to heal a shattered relationship or anger with God, either. I really thought that God had turned his back on me, but I was wrong, as you will discover when you read my story.

Certain people showed up in my life at just the right times and walked with me down an extraordinarily painful path. They helped me and they prayed when I could not. One of those people came from a totally different culture and walk of life than I. How I met him and how he helped me is a miraculous story by itself.

I am not saying stop your medications or give up on your physicians, I am not saying that at all! But I do want to express that there are other ways to heal. I am saying that no matter which way you choose, conventional or alternative or a combination (always consult with your M.D.), this is not an easy walk. Neither conventional, alternative, nor a combination of choices holds ALL the answers, either. You must take responsibility for getting well yourself, difficult though it may be. Realize and accept that healing from this kind of injury and putting your life back together is probably the hardest and most time-consuming work you will ever do in your whole life.

Be forewarned, too, that you, your brain, your life, your personality, your health, your relationships and your spirituality will not be the same as they were prior to your injury. No one comes out of this exactly the same. Many difficult, dark years may pass before you are healed. You could call your attempts a rebirth of sorts. Dare I say that you will be better than before your injury? I believe I am.

I function fully and am back working as a registered nurse. Numerous other accomplishments have brightened my days since the injury: I worked in a very demanding physician's office as an R.N., assisted the doctor, educated patients and also performed allergy testing. All of this is very responsible and exacting work. I resumed working in home care nursing, and act as nurse supervisor also. I especially love working with our TBI clients. I continue on with my practice of traditional nursing.

During my earlier phases of recovery, I completed a diploma program in homeopathy and also completed my post-graduate studies in that discipline. I currently tutor for a homeopathic school, and revised two of their distance learning courses. As you can imagine, my brain works well. I am content to continue tutoring for this school. I enjoy communicating with and learning from other international health professionals who view and handle health challenges in a variety of ways. I look forward to the day when there is equal freedom in personal health choices, not dependent upon the state or country you live in. However, laws must be respected as they currently exist.

There are many ways to achieve health. As my wise American Indian brother told me, you have to go with your beliefs, and that is what will work for you, whether that is conventional medicine, traditional native medicine or any other forms. Discuss those beliefs with your doctor. You do need a physician on your side. Find one that understands and respects your beliefs. The spirit-mind-body connection is all-important.

Recently, I helped teach at a TBI seminar to a group of 50 TBI Medicaid waiver providers in NY State. I wrote and presented a composite case study to demonstrate the total picture of (M)TBI to those providers. Their response, frankly, astounded me and I knew I needed to write about my own TBI experience. Education still needs to be done on (M)TBI. Many of the injured never get diagnosed. Most of them are misunderstood. Nobody really is grasping the full picture of what happens to somebody with this injury and what is happening with their "interior life." There is little appreciation of the extent of problems that reach into their relationships and family life. Few can comprehend struggles with an evolving "self." This is no small wonder. Unless you live this yourself, you could not have any idea.

One very big hurdle with a brain injury is the injured person's inability to explain what they feel like or what they are experiencing. They cannot communicate well, if they can speak at all. The medical word for this is "aphasia." I was aphasic and it's horrible. In this book, I hope to be a voice for those people who cannot communicate well, and who struggle with this dilemma. I wish to offer encouragement and hope to those who have a brain injury. You can get better. You can heal on the deepest levels. I want to lavish encouragement upon those who help this group of people. Don't give up. Have lots of patience, because this is going to take many years of hard work. Ultimately, (M)TBI survivors are like the phoenix rising from the ashes.

The composite case study I wrote and read for a recent NY State TBI conference resulted in such an impact on my audience that I need to share it with you here before presenting my own story. Painting a picture of an (M)TBI is helpful because, unless you live out that list of (M)TBI symptoms personally, it's hard to imagine what a person really goes through, not to mention the crushing, burdensome impact the situation exerts on the family.

TBI - A Composite Case Study

The victim of a minor automobile accident, the 45-year-old woman is released from the hospital. Married with two children, ages eight and 18, her devoted behavior now seems vastly different from what the family is used to. She sits for hours at a time without speaking or doing anything, unable even to ready the eight-year-old for school and be on time for the bus. She seems unable to cope with notes from school announcing upcoming events. She cannot plan things or follow a plan organized for her. Coping with grocery shopping is impossible. She appears oblivious to time and seems to have forgotten that she -- or anybody else -- needs to eat. Clearly, this woman prefers to go off to a dark room to lie down in silence, rather than be with family and watch TV or be in conversation. She does not interact much with family or with anyone else.

Extremely angry one minute, she can be very sad or extremely fearful the next. She is, however, very aware of this and tries to hold it all in so she doesn't appear to be a lunatic. She has also developed periods of anxiety which are worsening. Sometimes she exhibits rage in frustration over her condition. Her doctor diagnosed her with PTSD (post traumatic stress disorder) and a (M)TBI, but no family conferences are held, and the family receives no education about the long and troubled days ahead.

She begins using four-letter words that she never before used. She is having great difficulty remembering what is said to her or what she has said to others. She lacks the determination and grit with which she used to inspire and direct her own children. In return, they disobey, ignore her and laugh, saying "Mom's crazy." She either forgets to set the alarm to get up in the morning or sleeps through its ear-splitting sounds. Her own husband now accuses her of laziness, unable to understand why she cannot take care of things the way she used to. He thinks she's just searching for an abundance of sympathy and wishes she'd get over it. As time goes on, her pleas for help only get her arguments and refusals. She can't remember appointments because she no longer remembers to check the calendar. Becoming lost and easily confused is routine for her now. In the past, the family always celebrated holidays. Now she can't even comprehend planning anymore, so the children wonder why they are being punished with no holiday celebrations.

Sadly, no signs of affection are seen from mom toward her kids since the car accident. Her husband has taken on an increasing amount of work and hardly stays home now. She feels extremely angry at his absence because he is not there to help or spend time with her. Mom no longer makes dinner, so the kids fend for themselves. Ironically, they must remind her to eat because she forgets and just has no appetite. She has lost 15 lbs. in less than one month since that car accident. The 18-year-old has taken over the responsibility of putting her younger sister to bed and keeps an eye out for mom as well. What used to be normal bed and bath times are not observed anymore. Lacking the previous family structure, the kids simply go off to their own rooms and everybody is isolated in that house. Their mother cannot handle family time as it once was observed, nor can she explain its absence to them. She has neither the strength to enforce or even remember the old routines. She cannot keep up and the children end up doing what they want. Of course, they dearly miss mom and dad and who they all used to be. Mom is up and down emotionally and none of it makes sense. She is not who she used to be. Arguments between mom and dad are intense when he is home. The family structure is disintegrating.

The eight-year-old would like to go out with friends on weekends, but mom is unable to drive the car or make arrangements on the phone. Mom can't handle the shopping anymore either, so the 18-year-old assumes that duty. The kids don't feel that mom cares anymore and mom realizes she doesn't really feel emotionally connected to her kids and family in the same way anymore, and that burdens her with guilt. She feels she does the best she can, even though she realizes her best is not good enough. She knows only too well that her husband and children don't understand because she doesn't understand, either.

Time goes by, but mom continues depressed and strange-acting. She knows she's different from before that car accident, but just cannot reconnect to who she used to be. She cannot discuss things about which she used to excel. She's having a hard time explaining anything. She knows what she wants to say, but cannot connect the words to get it out of her mouth. Communication is a very slow process and very incomplete. The daze she lives in overwhelms everything.

What most people wouldn't realize is that it takes monumental energy from her to listen and to then understand what people are saying. If she is asked something, she has to remember the question and hold that information in her brain while she thinks about the answer, which she must hold in her mind also. It's a complicated mental juggling act. Then she struggles along to find the words she needs to answer with. And -- holding all of that together -- she must get the words out of her mouth. This takes an incredible amount of energy and time and it is mentally exhausting. In the interval, she gets interrupted by her questioner, even before she can get words out of her mouth. He changes the subject and asks an unrelated question and this alteration in conversation erases everything and leaves her with a blank mind. She has to start the process all over again for a new topic.

This pattern describes all of her communication, but what is especially frustrating is when a hurried doctor does this. He will cut her off, thinking she is done with her answer. He says something to her that she cannot understand and will not remember. Then he leaves the room to go on to his next patient and the doctor's appointment that she hoped could explain her actions or settle her mind is over. Frustration and anger build. Being given written notes helps, but she doesn't comprehend meanings behind the words and forgets what they relate to. Further, she forgets even that she has these written notes five minutes out of the doctor's office.

She can't remember anything. She used to love reading, but that's out of the question now. She can't remember what the first part of the sentence was by the time she gets to the end of the sentence. Watching TV and listening to the radio hurts her brain. There's not much she can do. Sometimes listening to music helps if her brain isn't fatigued. Otherwise, it's just an added stressor. Darkness with no sound and being alone is best when things are like this. Sounds, lights, people moving quickly, more than one person talking, enduring racket from the TV, any background noises, even from the street -- these are not tolerated well at all. She can only do one thing at a time and that requires great concentration. She can't work at her former professional job anymore.

She is scared and tends to panic because she can't remember simple and normal things. She used to be very independent and now is very dependent. Her kids now take care of her, reminding her to eat, to get up and to do a host of things she once routinely accomplished. There is no hope offered by the doctors. They advise that she has a concussion and no one can guarantee anything. They say it will take at least two years for her brain to "settle" and then they'll know which deficits are permanent. She knows she does not want to live the rest of her life like this, and if things get worse -- what then? No one understands and no one helps. She is referred from the hospital ER to her own family doctor, then to a chiro-practor and orthopedist. She is referred to a physiatrist and a therapist for talk therapy. The insurance company calls persistently, forcing her to see their doctors. They send her a rehab specialist who comes to her door without calling first and steps right into her home. She is very angry that a new self-appointed group of "professionals" is now trying to run her life, trying to absorb intimately the details of who she is and her relationships. She wishes they would all drop dead. Nobody asks her what she wants in her life. No one asks her what living this way is like for her. No one waits for her answer.

Her personality is unknown to even she at this time. Her brain works, but very differently. She has to decide how to use both because they seem so new and unknown. She makes conscious decisions on what she can do and be and what she cannot do and be. Most of it is a discovery process, trying out some things to see if and how they will work out. She will be the one who decides, not one of those other "professionals." Some of the professionals have remarked that she has a lot of anger. She thinks, if they could also visit this hell she lives in, they would think she had the patience of a saint and a ton of courage to keep on living. They tell her to get rid of the anger, but she knows the anger is helping to keep her alive.

When she does hold in the anger, things get black; she gets very depressed and wishes to die. She feels between a rock and a hard place.

Friends…? No, she can't deal with having friends over. They look at her strangely and their visits tire her out too much and leave her feeling even more confused and depressed. One friend told her she can't deal with her anymore, and in the blink of an eye the friendship was over. Friends ultimately stay away -- they don't understand why she no longer calls or joins them anymore. They are uncomfortable around her. As time goes on, friends get caught up with their own lives and move on without her. Mom vows over and over in the presence of the kids how much she wants to be dead, that she feels trapped in a black bottomless pit. Her personality is vastly different and she is more aware of that than anyone. She says repeatedly that she is not the same person anymore. She and her husband are having serious problems with the marriage and she is threatening to leave. The kids listen to it all and they, too, are depressed.

She develops severe dizzy spells, preventing her from walking and driving. She can only sleep two hours at a time and is wide awake the rest of the night. She can't dream anymore, either. She feels on edge a lot, as if she wants to jump out of her skin. She feels severe knife-like pain in her brain that wakes her up when she is asleep, and then the severe dizziness takes over. Among the sleep deficit, the stress, and the impaired memory and altered perceptions, she doesn't know if she said something or some-one else said it, or if she may have dreamt what was going on. The kids are very scared and depressed because they have never seen their mother like this before.

There are no supports in place for this woman or her family. "What can be done to help this woman and her family?" I asked my audience.

My Injury

The day of January 15, 2000 started out normally enough. As a visiting nurse, I was waiting for traffic to clear so I could make a left-hand turn into my patient's driveway. Suddenly, my car was struck from behind by another auto. By all appearances this was a "fender-bender," commonly interpreted as a low-impact, rear-end collision. But these familiar, seemingly benign and misleading terms quickly translate into acceleration/deceleration forces, a coup/contrecoup force upon the brain. My head and neck received quite a jolt from that impact, causing the brain to literally bang up against the inside of the bony skull. The immediate consequences, hidden from obvious view, involved swelling and bruises -- or in other words, concussion and brain injury.

I suffered no loss of consciousness at the scene, nor did I hit my head on anything. I did not go into a coma. After the other car impacted mine, my head flipped back and forth and sideways. I did feel as though things suddenly became very unreal, and I seemed to be in a daze. It was not long after -- mere seconds or maybe a minute -- that severe nausea, dizziness and light-headedness overwhelmed me. Definitely, I no longer felt the same as before the collision.

This deadened, anesthetic-like feeling inside my head became more pronounced as time went on. It crept up the back of my neck and over my head and spread out. My neck and head hurt. The intense nausea, light-headedness, fatigue and dazed feeling would remain with me, not for days or a few weeks, but for many months, well over a year later.

Doctors

I never saw my patient that day. Instead, I was taken to the hospital emergency room. In addition to the nausea, numbness and dazed feeling I was extremely sleepy, too. The ER staff did not order an x-ray. The examining M.D. only handed me a list of signs and symptoms of head injury to contact a doctor about. When I informed him I was well on my way already with two of those criteria -- the vomiting and sleepiness -- his response was, "Well, if it gets worse, go see somebody." He released me to go home and instructed me to follow up with my family physician.

My primary physician diagnosed me with whiplash and x-rays were taken of my neck. He said there was nothing much seen on the x-ray. "Go home and rest." he suggested. "You'll be fine in two weeks." I was not fine later, however. My doctor then referred me to a chiropractor.

The medical doctors had reported seeing nothing abnormal on my x-ray, but my chiropractor, Dr. Pete, saw problems with a bulging disc. He designated the problem areas by writing on the x-ray to alert my physician. I then took the x-ray back to my M.D., who loudly protested his displeasure that a chiropractor wrote on his x-ray. The doctor continued to lecture me that the chiropractors' profession had no basis in science.

I continued seeing both my M.D. and the chiropractor. My neck pain, and also the tingling and numbness running down my arm and into my hand, as well as my legs, were greatly relieved with chiropractic adjustments, and for that I was happy. But the excruciating neck and occasional arm pain, tingling and inevitable loss of feeling in my fingers always returned. I was going three times weekly for chiropractic adjustments.

Unfortunately, the pain did not resolve in the manner or time frame that even the chiropractor wanted. In addition, I was having problems remembering. I could not even remember my phone number, and this frightened me too, on top of the neck pain and loss of feeling that were my constant companions.

When I notified my M.D., he ordered a CAT scan, which showed no subdural hematoma. I mentioned to him, almost in dread of his reaction, that the chiropractor felt my symptoms of the still-present nausea, memory problems and dazed feeling could be a TBI -- in addition to the neck injury. My M.D., predictably at this point, stated loudly that chiropractic is not evidence-based, scientific medicine. (Unknown to me at that time, my chiropractor had focused his post-graduate chiropractic studies in brain trauma related to cervical spine injuries.)

Eventually, the chiropractor ordered an MRI of my neck because I was not improving as expected. Actually, my symptoms were worsening. This information infuriated my M.D., who yelled at me in his office: "What's this? Some chiropractor is ordering an MRI? A $1500 test, which the insurance company has to pay for? That's only for excruciating pain! There's a nursing shortage and I need my patients seen! When are you getting back to work?!"

As a visiting nurse I had visited many of my doctor's patients, as well as the other physicians' patients in this community. Had my eight-year-old had not been with me that day, I would have yelled back at him. How and when did he decide my pain wasn't worth investigating? I am fine to visit his patients, but my own health and pain mean nothing? But since my daughter was present, I spoke not a word.

The MRI was performed and showed moderate to severe spinal stenosis with a herniated disc at the C5-6 level. The injury had clearly pushed one of my cervical discs onto the spinal cord. This is a profoundly dangerous injury which could easily cause a quadriplegic (paralysis) state and could also require a ventilator for breathing if that disc totally compressed my spinal cord. Now I was truly scared for my life, for I had worked with many a quadriplegic and paraplegic patient over the years.

The next shock occurred when I hand-delivered this chiropractor-ordered, official MRI report to my personal physician's office: Dead silence. He never called or made any contact with me. There was no response at all from his office. That was the last time I ever returned to the unhelpful office of my M.D., and the only other time I would contact his office was to have my and my family's medical records removed. I was hurt and disgusted with the responses and, particularly, with the lack of responses I received when I needed my medical doctor's understanding more than ever.

My primary physician missed the major parts of my case, the things that put my very life in jeopardy. In addition, he told me I had better "do something with my anger" when I expressed anger at the man who drove his car into mine. He, however, was infuriated many times over about the chiropractor's care and concern about my case. In addition to all the major points of my case that the M.D. missed, it upset me that he thought his expression of anger about the chiropractor was ok, but my anger over being hurt in a motor vehicle accident was not. I need to emphasize that this was the last time I or my family ever went back to that M.D.

Cognitive Problems

While still in his care, my M.D. did order a brain CAT scan for me when I told him I couldn't remember my phone number. That particular incident involved me giving information over the phone to my employer. They asked for my phone number and I gave them one, but a few seconds after I spoke I realized to my horror that it was incorrect. I said, "No, that's not right!" I attempted again to give it. I correctly spoke the first three digits of my phone number, but somehow combined it with that of my employer's. Realizing my second mistake, stunned and aware that I no longer knew my own phone number, I became terrified.

I remember well how I sat silently in my kitchen after hanging up the phone. I looked at all of the familiar things surrounding me, the refrigerator, stove, sink, phone, floor, ceiling, … everything. Each individual thing I comprehended and understood what purpose each served. All things together, though, in one room -- well, there just was no cohesiveness to it all. I couldn't tie it all together mentally to know I was in the kitchen. They were just isolated things unto themselves. There were no automatic associations coming into my mind for any-thing. I thought to myself at that moment -- is this the state a baby must be in? This would indeed be the state a baby must endure until it learns what these things are for. As experience and memory builds and holds the experiences, a complete picture forms and automatic associations take hold in the brain. A blank slate must be how we all start out as babies, I reasoned. I could think these things, but I could find no words to verbally express the thoughts. How frightening it was to think my mental capabilities had declined to this appalling level.

Diffuse thoughts.... that is what I had. My thoughts, and later on my speech, would begin with one topic and end up somewhere else that others did not recognize as "appropriate or related." In the medical world this is called "tangential thinking." I called it diffuse thinking. Everything is connected to – everything! The problem was that I understood and saw that, but no one else did. According to others, I lacked focus and did not answer questions. I thought it unfortunate they could not see the wonderful connection of these things I now understood. Tangential thinking is considered a "cognitive loss" by the experts, but for years and even now, I thought it was pure enlighten- ment and I enjoyed it. I was seeing the world from a different, con- stantly changing perspective. Personally, it was not a "negative" for me, but the rest of the world does view it as "dysfunctional."

My profession had been nursing for over 20 years when my in- jury occurred. I had cared for and seen many varieties of patients and conditions. Never in my life could I have guessed this was what it was like for someone with a concussion or a stroke. But I was suddenly living the excruciating lives of my own patients, unable to tell anyone about the terror and wonder of it all. I could not connect to the words stored in my brain to tell it, or to scream it out coherently. Let me tell you this: No matter the silence, the awareness of a brain-injured person can surpass that of an uninjured person. It is just a different kind of awareness.

My brain CAT scan did read as "normal" and my doctor was re- lieved that it did not show a subdural hematoma. However, the scan showed no reason why I couldn't remember my phone number, could not comprehend the totality or implications of common items found in a room, and couldn't untangle a maze of meaningless words to express myself coherently.

Later on, when words started to come, occasionally, I would use the wrong ones. For example, one time I wanted to say the word, "cat." Instead, the name of a local grocery store is what came from my mouth. I was shocked. This was happening often. I learned to speak very quietly so that when this would happen it would be less noticeable, if noticed at all, by others. But still, I was happy to have words to speak, even if sometimes they were the wrong ones. I concentrated on what I wanted to say, and rehearsed it to myself before saying it out loud.

I Am Not The Same

"I am not the same."

That was all I could say to anybody. I knew I was not acting as I used to. I could not think and converse the way I did before, and I couldn't help any of it. I could not articulate all these new feelings and sensations that began dominating my life. I was in considerable distress from the neck pain and the cognitive and emotional changes I had noticed about myself. Especially frustrating was the inability to articulate all of this.

I thought maybe the pain was adding to this cognitive stuff I experienced. I was very fearful about what all of this meant, and I remained silent for the most part because I could not connect to words. Inside my mind, though, all kinds of ideas and emotions played out intensely, bouncing all about, raging and fleeting. I had great difficulty focusing on just one subject when I tried to think. Fear and anger interrupted every idea in my mind. When I tried to hold down the anger and rage, I felt very depressed and couldn't remember anything. I started using four-letter words that I had never used before in my life and I was shocked at myself. Where was all of this stuff coming from?

The strange thing was that I didn't see anything wrong with it. Those words helped to express what I was living and feeling. I existed in a different reality now, as if my mind had moved to a new place, a strange place I never imagined. After raging at someone out of frustration and anger, I would settle down, only for the recollection of what I said to the person to emerge later in my consciousness. Then I would be horribly embarrassed and worried that I had damaged a relationship or hurt another person. I was shocked, amazed, at how doing this was just as easy as breathing and not even noticed until later.

My chiropractor, Dr. Pete, God bless him, withstood lots of this from me. Each time he would say, "Tomorrow is going to be better." When I apologized to him he would always say, "There is no need to apologize. We move forward and we don't look back." He was the first to tell me he suspected I had a brain injury -- and that this injury was the reason why these bizarre things were happening. I cannot tell you what his kind, compassionate responses meant to me. He became a very good friend when there was nobody else. I was especially impressed with him because he had not known me at all before the car accident.

I was very happy with my chiropractor's care, but the truth is that a chiropractor's care and diagnosis is not taken seriously by the experts in the medical field -- and takes a pathetic back seat to their opinion once the legal system becomes involved.

Workers Compensation

My injury occurred while I was working, therefore I was thrown into the Workers Compensation System -- whether I wanted that or not. Do you know it is "illegal" to use your regular health insurance for an injury incurred while working? It is here in NY State. It is also illegal to pay a medical professional out of pocket. All of your medical care has to go through the Workers Comp system if you are hurt on the job. To complicate matters, there are many physicians who will not take Workers Comp patients. Usually, they are exactly the physicians you would like to see.

I understand why some doctors will have nothing to do with this system. But I especially appreciate those who do put up with it for the sake of people like me. They are lucky if they get paid for their services in a timely manner. If summoned to court to testify, they are paid neither for their time, nor for patient appointments they can't schedule. The injured person also has a lot stacked up against them in this deal that they have no control over.

So -- just who really runs the system? The insurance company and big business really have the upper hand. If a physician asks for a written letter guaranteeing payment from the insurance company, and they won't provide one, you simply won't be allowed to see that physician. The same situation happened with a neuropsychologist to whom I was later referred, and I never got to see him. Basically, the insurance company ignored the request for the letter guaranteeing payment.

If you enlist an attorney's help, you have more leverage because your attorney can call and request again -- after waiting 30 days. Because of all the paperwork and bureaucracy, you will wait yet another 30 days after the second request. If the insurance company still does not comply, then your attorney has to request a hearing with the Workers Comp board. You have now waited 60 days for medical care. You wait at least a few more months, if not longer, to get a hearing with the Workers Comp judge. As the hearing proceeds, the insurance company offers an excuse and an apology to the judge, assuring the judge they will comply.

Now you're back to waiting for that essential letter. When the physician's office receives that, you'll need to schedule an appointment, and in many instances that will be another frustrating month's wait, at least a month's wait. Do you see how long you have to wait for medical care under this system if those with power want to prevent you from having access to it?

One of my own physicians referred me to a neuropsychologist. By the time it was formally agreed upon it was way past the time frame when it would have been helpful. As my new M.D. said to me, "It's too little, too late now. These things need to be done in a timely manner - not nine months or later." Keep in mind the fact that it is illegal to pay out of pocket or to use your regular health insurance to pay a physician for care when you are hurt on the job. This is how the system exists in this country and allows workers to be treated -- citizens who pay their taxes on time and faithfully support their communities.

Here I was, experiencing constant physical pain, aware that I was acutely at risk for quadriplegia and paralysis, as well as requiring the dreaded ventilator. I had a brain injury, and couldn't even get my own phone number straight, I couldn't hold my head up with the pain, I couldn't read and comprehend or remember, I couldn't watch TV, and I couldn't think. I could only be minimally comfortable lying down. Sleep lasted two hours, maximum. I was up most of the night. I couldn't relate how I was feeling to anyone, either, because I could not connect to words to tell anyone anything. I could only feel extreme emotions as my mind flooded with fearful thoughts. Simultaneously, I was very aware of all that transpired around me. A TBI can make you feel dazed, but also it can place you on hyper-alert. People think because you have a brain injury you do not notice what is happening or will not remember. That is not always so. Just because I could not remember my phone number did not mean I forgot the kindnesses -- or the cruelties. Those things stay in your mind and do not leave. Those actions leave a mark on you. Those little incidents you remember. Also, because you have a brain injury, some think that you are stupid. Nothing could be further from the truth. A brain-injured person's intelligence will amaze you, as will their inner strength in just choosing to be alive and striving to exist in the condition they are dealt. But they are not able to advocate or deal effectively for themselves with such a complex system as Workers Comp. I could not even carry on a conversation with my attorney when I finally had one. Neither could I comprehend or remember much of anything except direct, simple statements that he would constantly repeat to me as I asked the same questions over and over.

Many weeks had gone by and I did not improve. I worsened in terms of cognition, emotions and with day-to-day functioning. In fact, I was barely functioning. My employer's disability carrier, the insurance company and my family physician were getting very frustrated with me because I was not back to work. As things stood, I was lucky if I could get dressed, remember to eat and keep my appointments. That's all my days consisted of.

On the other hand, my chiropractor supported me with encouragement, good thoughts and words while I visited. Most of the time, I was cognitively unable to hold a discussion with him. In the meantime, there was no family time at home to speak of. I stayed in my dark room and my kids functioned by themselves while my husband worked. There were no dinner times, no pleasant remnants of the family life we all had known, and everyone fended for themselves, day after bleak day.

Reign Of Terror

The United States, in response to the 9/11 air assault and other ter-
rorist acts that changed our lives in so many ways, conducts what is commonly known as "the war on terror." Not to be melodramatic, but there is
also another form of terror -- a perfectly acceptable terror -- lurking
silently, ready to attack at the mere sight of a medical emergency. On the
prowl constantly for its next victim, this terrorism has a name and it stands
hand-in-hand with all the power of a legal system which, supposedly, works
for all of us. Its name is *insurance company terror.*

The "terror" occurs when an insurance company declines to pay
medical claims or disability. Nobody talks about the harassment or the
degradation tactics they so expertly employ, either. I wonder how in the
world anyone can even look in a mirror when their work primarily involves
making threatening and harassing phone calls to someone who is sick or
hurt. What sort of people does that for a living? The insurance companies
excel with people like that on their payroll and, make no mistake, this brand
of terror is legal in this country. I endured a living hell at the hands of
insurance company representatives. I hold them primarily responsible for
my diagnosis of PTSD (post traumatic stress disorder).

As a nurse, I'm well-versed in being non-judgmental, being empa-
thetic and serving as a patient advocate. At the very least, I am good to my
patients. I've always prided myself on that, as well as the hands-on nursing
skills and assessments I've practiced. My work has always been praised, and
always described as thorough. In most instances of employment, I was

invited to work - I did not gain employment because I applied for positions.

However, after I was seriously injured and could not work, I received nothing but insults and threats, insinuations that I was a fraud and malingerer from those representing my employer's insurance carrier, when they judged that my time was up. How do we allow such people without any medical training or background to judge another person, much less their injuries and illnesses? When it comes to saving money, the rules favor the insurance company. The cruelest maneuvers often seem to be reserved for those who are injured and cannot defend themselves, in an attempt to save the insurance company's money. How does someone with a serious injury defend themself against this barrage of threats and medical claim denials when they cannot verbalize, cannot fully comprehend or remember, and cannot even get eight hours of sleep at night because of a brain injury?

I continued treatments with my chiropractor, and the insurance company continued berating me with their annoying phone calls, interrogating me and demanding to know why I was not working. Their harassing phone calls added greatly to my stress and distress. During this time, the pain continued non-stop and I was feeling very overwhelmed, scared over the prospects of my herniated disc and compressed spinal cord, and greatly concerned that I could not think, express myself at all or remember anything. The insurance company's response was a threat to not pay my medical bills unless I signed their ever-present blank consent forms (later on they made good on their threat, even when I did sign). Their harassing phone calls continued routinely. I ranted to Dr. Pete one day in his office that, if nothing else, I was proud that I never in my life, EVER, treated another human being as wickedly as the insurance people treated me. The worst I could wish on them would be to receive the same kind of injury that crushed my life, and to have people just like they manipulate their case in the same way they did mine. I could hardly believe such people existed. It seemed to me as if they were full of hate, as if client/patient hatred was a prerequisite for their job.

Over the course of my recovery I was forced to see two IMEs (independent medical examiners). These are, in a few apt words, merely physicians bought and paid for by the insurance company. Dealing with the IMEs was an incredible experience, especially because I'm a nurse. I know how a patient should be handled and treated. I know what constitutes poor treatment. These doctors never even referred to me by my name. I was called "you." Prior to the examination, a stained patient gown was offered as my exam attire, and then I was directed into a dirty exam room still adorned with a crumpled, used paper sheet from the previous person's body on the table. I was told to sit on that table without any of the trash being removed. My reflexes were tested with the back of some bandage scissors by yet another IME -- an orthopedist. I could not get over the unprofessional manner embraced in these professional activities.

For a time over the course of my recovery, Dr. Pete and Dr. Abby -- who would be my new family M.D. later on -- and my therapist, Mary, were not paid by my employer's liability insurance company. Never in my life had I gone to a physician and not paid. I am the type of person who would never go to someone if I couldn't pay them as I have too much pride. But now, I was not allowed to pay out of pocket or to use my regular health insurance either. This added greatly to my stress level. Those who were helping me and did the most for me were not receiving their pay, and it became clear that they, too, were being punished by the insurance company. Because they were such good people, all three would say things like, "Don't worry about it, let's just keep on and get you better." I have nothing but praise for those who helped me. I have nothing but admiration for doctors who help people who have been injured while working.

On good days, I would get up and try to function in my deeply dazed state of being. On those good days, I would also try to go for a walk. I always felt better outside, walking in the fresh air and seeing trees and plants, even the snow. I composed simple haiku about nature. I thought it a good cognitive exercise. I was trying hard to get better. I always managed to eat a good diet my whole life. I continued on with only quality whole foods when I did eat. My neck was slowly getting better, but I still raged with anger inside and that alternated with depression. Half the time, I did not want to eat. I couldn't carry on a normal conversation or discuss anything. It's impossible when you have no spontaneous thoughts, you can't connect to words, and all you feel is intense emotions.

My friends, though very concerned, gave me a hard time. One did not want me going to a chiropractor because of the neck injury's severity. Therefore, to assure her and others, and hoping to satisfy the insurance company that my injury was legitimate, I agreed to see an orthopedist.

"Dr. Dread" did an exam, looked at the MRI and x-rays and discussed the ways this could be treated. I latched onto his key words: "Conservative treatment, continue with the chiropractor, and come back in one month." He asked if I wanted pain pills and I said no. I would stick with the adjustments. No sense drugging myself if other methods worked. The last thing I needed was to become dependent upon pain pills. The other last thing I needed was another medication allergy. As a nurse, I knew that more damage might result to an injured part if you do too much.

Overdoing is likely to occur if, for instance, you can't feel pain because you are taking pain meds. Also, how would I know if I was getting better if I could not assess my own pain level -- if I was out of touch with it? When a part is truly healing, you have less pain, and that's a sign of healing. When you are healed, there is no pain.

I knew that if I began the pain med regimen the insurance company would announce that I was "cured" and give me a hard time if any complications arose later. I did not need pain meds before this injury and I certainly anticipated a return to pre-injury status without them. I also harbored substantial concerns that a pain medication's effect on my impaired brain function was not going to improve it. Under no circumstances did I want a pain medication. All of these things were on my mind, but I was unable to communicate these thoughts to anyone effectively.

Much later, when I returned to the orthopedist, he actually seemed upset that my pain level was much better from chiropractic treatments. He told me I would end up a quadriplegic if I did not let him operate. Knowing full well that he could not guarantee a successful outcome and cognizant that surgery could also leave me paralyzed, I told him, "No, because I do not want to risk becoming a quadriplegic." His response? He got angry and yelled, "Get out and don't come back unless you agree to surgery!" I couldn't believe what I was hearing! I used to visit this orthopedist's patients for him in the past. Now I was out both an orthopedist and a family doctor. All I had was my chiropractor. The only person who could recognize and take seriously my injuries was the same person at the bottom of the system's totem pole.

The harassment and phone calls from the insurance company just would not quit. They sent letters and they left messages on the answering machine. They determined, based on their "formula," that my disability was worth $40.00 a week in compensation. My husband was furious and argued with them on the phone.

I could not work, I had a severe neck injury, and I could potentially be a quadriplegic. My cognitive abilities were gone. I also had no family life, as I was not able to take care of anything or anyone in the house. My kids raised and took care of themselves and they looked after me. My kids had to remind me to eat and even to get dressed, or to sit with other people instead of just lying in bed. A lot of my time was also spent in cervical traction, something I did at home by myself. I was very concerned that I could neither understand nor deal with notes from school coming home with the children, announcing upcoming events. I was unable to plan anything for my kids, unable to let them have their friends over. I lacked the ability to take them to their friends' houses. Nor could I even cook a meal, let alone plan one. My energy was drained and I sat, walked or lay down with a blank mind and full of fear the whole time.

That I had been a top-level functioning medical professional of many years prior to this injury didn't matter to the insurance company. They also didn't care that I was not back to pre-injury status, either. Their only driving concern was that I earned, at a minimum, $40.00 a week in any capacity. If I was able to do that, they were off the hook in paying any compensation. They were covering my medical bills -- well, the ones they decided would be paid. The previous year, I mostly worked weekends. Nurses don't earn big money, so my disability was based on a formula that designated $40.00 a week in disability compensation. Eventually, because of this, I had a very difficult time finding a lawyer because his pay would be determined by my rate of compensation. One attorney actually laughed at me and said he considered mine to be a charity case.

Anger and fear over the insurance company's harassment techniques continued to overwhelm me, in addition to the fact that my original family M.D. wasn't concerned about the state of my health. I was very worried that I could not think or remember. I could not even watch TV because I could not follow any story line. I could not read because I could not remember what I was reading or comprehend it. I couldn't stand to be around other people because the sounds of their voices overpowered my mind. I could not handle being in a room with lights on and with more than one person talking to me. I could not have TV or music on and have someone talk to me simultaneously. On some days, even noises from the street seemed too much. I was exhausted. I could not sleep more than two to three hours at a time.

And I was fed up with other people telling me what to do with my neck and whom I should see about it. The terror that I may never be ok again engulfed me as a matter of course. I loved my job and I wanted to get back to work, but there was no way I could go out and be responsible for a patient when I couldn't take care of myself. I missed my friends dearly, but I couldn't take their criticism over my coping actions. They were too critical and too harsh. Their words hurt me. I knew they were trying to help, but they only made me feel much worse. In my opinion, friends should be supportive, not judgmental. Their visits left me confused and depressed. One of my longtime friendships suddenly ended over this injury.

Friends and family were telling me I had to pull myself together. They insisted that my perspective of everything was all out of proportion.

Their words not only hurt, but infuriated me. I was unable to connect to words that might explain. I just shook with anger, fear and frustration instead. I felt under such intense pressure that I had never before experienced. My husband took on more hours at work and his absence increased my fear of being alone. In response, I went to my room and stayed in the dark. I absolutely didn't want my kids to see me this way. I was overflowing with stress and felt as though I was really going to lose it. My husband could not understand my inability to handle anything. He and my friends wanted me to "get over it." The nausea, dizziness, dazed feelings and impossibility to focus on anything persisted this whole time. I especially felt bad that I could not connect to my kids emotionally. I felt guilty because I could not do anything for them. I kept thinking that if I got through this and came out of it ok, I still would have missed lots of quality time in their lives. I knew more than anyone that I was different and that I was not the same, but I just could not connect to whom I used to be. I could not even feel emotions in the way I used to.

My perspective on life changed dramatically from day to day. Nothing was stable in this surreal atmosphere except for feelings of intense betrayal, fear, anger, sadness, rage and depression. I felt betrayed beyond belief -- betrayal by my physician, betrayal by a friend who insisted our friendship was her way or not at all, and betrayal by my own husband who chose work over spending time with me. I also felt much betrayed by God. (I am giving you my perceptions at the time. I realize now there was no way they could have known what I was going through.)

One of my friends did call my homeopath, who lived in a different city, to let him know I needed help. He called me and then put a remedy in the mail to me. I took it and for the first time since the car accident I felt all the tension inside myself unwind. I was able to sleep and cry alternately. It made me feel so much better to release that tension and to sleep.

Spirituality and religion had always been important in my life. I had been praying for help since the injury, but events only turned from bad to worse. I was very angry with God because everyone was against me, He was of no help, and I could not even depend upon my own brain. I was extremely angry at God. I did not want to live as a quadriplegic. I had cared for many people as a nurse to the best of my ability. I had always given the best. Why did my own physician not care about my troubles, when I had visited so many of his patients to his total satisfaction? Why was there no help for me when I had helped so many in the past? I did not want to live as a quad or on a ventilator. I was terrified and I would rather be dead than terrified.

The words, "scared to death" are meant literally. Did you realize you could be so scared that you want to die? That was the place where I dwelled. If I had seen a door marked "Death" I would have gladly walked through it to escape the fear I was experiencing. I spent a great deal of time in a black, bottomless abyss inside my own mind. It was filled with pure, unending terror, and so was I. The abyss grew wider and deeper as time progressed. I wanted to be dead to escape it. I wanted nothing more than to be out of it. I said these things often and out loud. My kids heard this and it scared them. My husband and I were arguing more and more and I wanted out, but I knew I couldn't work or do anything to get out. None of this was good for the kids.

Home used to be a sanctuary and a place of peace. The day came when I could no longer feel safe inside my own house anymore. That day began with a visit from the "rehab specialist."

A rehab specialist, sent by the insurance company, came to my door one day. He came without calling or any prior notification. He rang my doorbell. When I opened the door he stepped right past me, to the inside of my house. I was alone that day. The kids were at school and my husband had left for work. He started out in a friendly manner, asking me to sign more consent forms. I refused, stating I had signed enough consent forms already. He immediately became angered and in a raised voice insisted that my medical bills would not be covered unless I cooperated. I could feel my anxiety level and blood pressure increasing greatly. I told him to get out. He would not go unless I signed the papers and ultimately I did sign them, and literally threw them at him to get rid of him. When he left I was visibly shaking. I truly felt as if I would either have a heart attack or a stroke, I was so extremely upset. The pain in my head was incredible and my heart raced frantically, worse than anything I had ever experienced. I am truly amazed that I survived his surprise visit because of the effects it had on me. Don't ever underestimate the added stress of an upsetting situation or person -- and the effects upon someone who is already compromised.

A lawyer. I never needed a lawyer before in my life! Now I was being told the only way to stop the harassment of the insurance company was to get a lawyer. My husband handed me a phone book to look one up. I was incapable. All that print, all those letters and words on those pages -- I could not focus on any one of them visually or mentally. None of the words held any meaning for me. I had extreme difficulty dialing a phone number correctly. I could not talk well enough to explain what I needed. I could not listen fast enough, or comprehend what someone said to me, or write things down fast enough to even try to use notes as a reminder. All I could do was ask for help. However, sometimes asking for help was met with an argument and a refusal.

Now, the thing about most people with (M)TBI is that they look great from the outside, even though serious damage takes its toll inside. I happened to be one of those people who looked in good shape physically, so it was difficult for people to understand that I had some serious underlying problems. All they could see were behavior and emotional changes. I wanted to be in quiet areas, I was not talking much, I was in pain, I displayed a lot of emotion, I could do practically nothing, and I preferred to be alone. They expressed deep concern when I started saying that I wanted to be dead. That's all that they saw from the outside, and that's all I could manage to get out of my mouth.

I was starting to feel even worse. I would experience episodes where it felt as though my body was shot through with adrenalin, as if I was plugged into an electrical outlet. I would get extremely afraid during those times and would shake. At the same time, I could not tolerate being spoken to, and I could take in no more information. I could not handle lights or sounds. I would have to lie down and just wait for this "attack" to pass. I used to say, "My brain is bottoming out." Years later I would learn from a discussion with a neuropsychologist that the medical terminology for this is "brain fatigue" and it goes right along with a TBI. These attacks seemed to be happening more and more with me. I was not able to explain it very well to anyone, either. The more active my day, the more it seemed these things would happen.

My chiropractor saw me sitting in a fetal position in his exam room in this acute "brain fatigued" highly-stressed state. I knew I needed a primary care physician, but my own wasn't any good, and I refused to go back to him. I had no idea where to go, and it wasn't as if I was in any shape to find another doctor. I couldn't even read names in a phone book. I needed immediate help. My chiropractor told me something that I professionally would have told anyone, as there is no other answer: "Go to the ER."

I said to him, "Send me to someone who knows more than you, not less!" I did not go to the ER, as I feared I would be labeled a psych case. Later on, I would find out that many people with (M)TBI are in fact labeled a psych case instead of being diagnosed correctly. I wonder how many have been and are placed in psychiatric institutions instead of getting the help they really need? The day arrived when this fear became so intense that I thought I might do something harmful to myself and not even realize it. I called a friend and asked her to help. She contacted a friend of hers who happened to be an M.D. I was informed she would see me right away. A friend drove me to this new physician, Dr. Abby. She kept her office open for me when she was just about to close for the rest of the day. She gladly stayed in the exam room with me for well over an hour. She did an exam, she talked with me, and she watched me and listened to me. She looked at my x-rays and MRI. She asked questions and watched how I answered, or attempted to answer. I told her about my "brain bottoming out" and what that was like. My friend, who accompanied me, looked horrified as she listened.

Near the end of this very long visit, Dr. Abby told me I suffered a brain injury from the car accident, and that I was not going crazy, that I was in an acute panic state, and that I attained post traumatic stress disorder, also. She told me all of this resulted from the head and neck injury. She patiently explained a little bit about the damage to my nervous system from the head and neck injuries. I was not happy at all to be given an abundance of diagnoses, but I was happy that finally there was someone who understood and who had an explanation for this hell I was living out.

Just having someone like her around meant everything. She knew how fearful I was that I might need surgery that could further affect my brain and spinal cord. She sympathized with my fear of ending up on a ventilator, paralyzed. She told me that if worse came to worse and I was running into problems with loss of feeling in my limbs or inability to move my limbs, she would hook me up with the best neurosurgeon on the East Coast. I related my depressing encounter with the orthopedist and she replied, "Absolutely no orthopedist, only a neurosurgeon, and the best one if you need to have surgery on your neck or brain."

Dr. Abby and I discussed how there are no guarantees, especially with this kind of surgery, and it was something I wholeheartedly wanted to avoid. I was able to move my arms and legs and breathe on my own, and did not want that jeopardized. I did not want further brain damage caused by anesthesia during surgery. She understood from working in the medical field that medical workers see the worst cases of all and, consequently, fear the worst for themselves and those they love.

Her advice was to go home, rest and not overdo. She supported my decision on how I wanted this injury handled. We talked a little bit about the brain injury. She said there really was nothing but time and rest to make this better. She said there were no guarantees what cognitive abilities would come back or not. She wanted me to see a neuropsychologist, but the insurance company ignored the neuropsychologist's request for a letter to guarantee payment. I did not get to see the neuropsychologist and I did not have the inner strength, the memory or the cognitive abilities to fight the insurance company by myself, either. I do not exaggerate when I say that absolutely all the energy within me was used to keep me alive. I could not risk putting any of it out for anyone else. These weaknesses in others translate into how insurance companies make and save their money. Who looks out for those who cannot speak for themselves? For those afflicted with TBIs, and the elderly? These two groups do not have a fighting chance when it comes to this kind of bureaucracy.

In my case, in addition to the harassing phone calls, there was the stalking. Yes, later the insurance company showed my attorney a video one of their employees had taken as he followed me around. The insurance companies are relentless. I don't know how this kind of treatment is allowed to be perpetrated upon those who are sick or injured. In the worst situations, people go without needed medical care because they are too weakened to fight such ruthless corporate entities. I was lucky that I was not the sole provider for my family and dependent on that $40.00 a week they determined was my disability compensation. Around the time of my injury, my husband came across an article published in a magazine about a man injured at work whose disability pay was held up. The insurance company was disputing his injury and they stalled his case off for many months. As he was the primary provider for his family and was unable to work, and with no help coming in, and his wife unable to earn what he had, he decided that his family would be better off if he were dead and no longer a burden. He committed suicide. The next day a disability payment arrived at his house.

Dr. Abby was a wonderful ray of sunshine who brightened away the clouds obstructing my mind's view. She was very good with simple explanations to which I could relate. She was respectful of how I wanted my injury cared for. She was particularly adept at holding her own against the insurance company as they harassed her office about me. They continued to bother and intimidate my chiropractor's office as well.

Dr. Abby refused to turn my medical records over to them when they presented her with their forced-signature consent forms. She instructed them that I needed to sign specific forms from her office before she would release my records. She was like an angel, phoning my home a few times to check on me, and she sent me a nice card, telling me she was happy to be my doctor. But the insurance company didn't share in that happiness. They refused to pay the medical bills, continued their harassment and I was at last forced to consult an attorney.

I don't remember how I obtained my attorney. I think somebody handed me a card with a Workers Comp attorney's name on it. I can remember going to meet him, however. My mind quickly became overwhelmed by the cars parked on the street and the cars moving in front of and to the sides of me. The traffic lights, the people crossing and walking – everything was too much information and it inundated my brain as it and I attempted to cope.

I was driving alone to the appointment on this rainy day and the dampness was making my neck ache even more. I was feeling shaky with fear from all the street action, sounds and lights. I couldn't remember the old familiar streets. By the time I found the lawyer's office I was in excruciating neck and brain pain. I was shown into his conference room. I shook his hand, sat down with my papers and laid my head on his table. He looked at me in amazement with his mouth gaping open. He asked me if I was alright. I said, "No, but I will be." When he requested various forms from me, I told him I had no idea what they were, but everything was there in the stack and to please help himself. He asked me the same thing again as I sat there with my head on his polished table. I responded the same way -- could he please get them himself because I just could not. He didn't realize I couldn't even recognize those papers, had I seen them. I couldn't focus on finding a word, or group of words, on a page of paper loaded with nothing but words.

I explained to the attorney how the rehab specialist came to my house and what he did. I told him about all the harassing phone calls and messages on my answering machine. The attorney advised strongly that I was not to get into any more conversations with, or to sign anything else for that insurance company. From now on, I was to give them his name and he would take care of them. (For a wonderful book to read on the Insurance Industry and the way it handles you when you are disabled, read: "Insult to Injury: Insurance, Fraud, and the Big Business of Bad Faith", by Ray Bourhis. In addition, Michael Moore's documentary, "Sicko" is quite an eye-opener!)

So now I had gained three people working on my side: The attorney, my new doctor (Abby) and my chiropractor, Dr. Pete. By the time I "found" Dr. Abby and she diagnosed me, five or six horrible months had elapsed since the car accident. Searching to find just these three people had been one long haul -- but not as long as it would take to recover from the injuries I sustained.

Ghosts

When my new physician, Dr. Abby, informed me I had a brain injury, besides cervical nerve and spinal cord compression, and post-traumatic stress disorder, she expressed dismay that she could not visualize my optic nerves on exam. I said to her, "I do not want a burr hole! I do not want an Omaya Reservoir, or decadron, or anything like that!" She took one long look at me and said, "I don't think we are going to need that, JoAnn, but we are using mannitol instead of decadron now for these situations." She smiled and I laughed. I replied to her that it had been quite a while since I had done hospice nursing, but back then it was decadron that was used for a swollen brain.

Dr. Abby continued, "I need you to go immediately to an eye doctor because I cannot visualize your optic nerves. But for some reason, I always have difficulty visualizing optic nerves. So, we'll let another professional do that, and that person can dilate your eyes if need be." Only a medical professional can appreciate the severity of this situation. I did go off to the eye doctor who was able to visualize my optic nerves, greatly relieving me of my fear of having a surgical hole drilled through my skull. Yes, there is such a thing as knowing too much.

All of this knowledge and information weighed heavily on me. Every night I saw my former quadriplegic patients in my mind. I was amazed how I could see every one of their faces and remember their particular circumstances. One in particular came to mind. She could only move her head from side to side and relied on home health aides to take care of her. She had lived at least 20 years like that. A nice, gracious person, she was overjoyed when her physician agreed to sign a DNR (Do Not Resuscitate) order. That order would protect her from CPR and resuscitation attempts, in the event she had a heart attack, or was dying from some terminal illness. I could easily understand where she was coming from. I remembered each quadriplegic and paraplegic and I was adamant -- I did not want to live like that!

I knew that my nocturnal "ghosts" were not abnormal. I had a nurse friend who, when she was diagnosed with breast cancer, told me that every mastectomy patient she had ever cared for came up in her mind at night for her, too.

Night after night, I was semi-aware that I would "pick" at my blankets and bed clothes. I would sit up automatically, up and down, half asleep, in some different level of consciousness foreign to me, during the few hours of sleep I could get, and then I'd lie back down. Part of me knew these behaviors were from a brain that was injured. It is from cells deprived of oxygen and from increasing intracranial pressure. I saw in my mind my patients, from years past in the ICU, who were brain-dead, but their bodies still breathing on ventilators. I knew the nitty gritty moment-to-moment details of their care. But, who do you tell when you have no words? Who do you tell when you cannot remember the words, but the diffused thoughts persist, and you exist in a perpetual post-concussive gray haze? Some things cannot be fixed. They must be accepted and lived through. I wondered if I would "wake up in a coma" or "wake up dead" or, worse — wake up a quadriplegic. They say physicians and nurses are the worst patients in the world. We are that for sure.

Losses

The life I lived now was no longer the same as before, not even remotely. My work and income were gone, socialization with my friends was gone, and daily familiar routines were gone. My feeling of mastery over simple things was a vacant space. Sleep, memory, comprehension, the ability to read and the enjoyment in reading had departed. I could not watch TV. I could not organize fun things for my kids, either. I could not plan meals. The concept of time meant nothing. Everything was just one continuous NOW. There was no beginning or end to things in the way I perceived them. My ability to dream during sleep was gone. There was no comfort even in what little sleep I got. I would awaken in a startle, as if someone had violently and unexpectedly shaken me.

My familiar stable emotional connection to those I loved was gone and any connection to my former "self" was gone. What used to be my life was shattered, all of it extracted in the blink of an eye in an auto accident. In its place grew a new self that had not totally formed yet and still was in the process of change. My new unpredictable brain caused a flood of overpowering emotions that sometimes made no sense to me. If anything was clear, it was the loss of my independence. My ability to walk was different because of overwhelming spells of dizziness that rarely announced their approach in advance. No one can promise you a return of those abilities most cherished. No one offers you any light at the end of the proverbial tunnel. The one thing always stressed to health care professionals by their teachers is not to offer a patient false hope. I can now say that offering hope is never false. Hope should be offered and it is not unrealistic. Any serious brain injury is too much to bear without any chance of hope. As long as the person with (M)TBI is motivated and attempting to get better, gains will be made, no matter how slight. No one should ever tell them, "No one knows if you will get better." What should be said and enforced is, "Try. We will work together and you will see gains, although it may take a very long time." That is not a lie. That is not offering false hope. That is offering realistic hope and mark my words, the person with (M)TBI is desperately in need of that encouragement. Understanding and support should also be given to someone who suffers so many losses. This is no different a grieving process than one experiences after someone close to you dies. In a way, this is worse, because nothing in your life is the same again and there is no closure, as there generally is after a death. Comfort cannot be taken in old familiar things or even persons, because you cannot perceive them in the old familiar ways.

Anyone taking care of a (M)TBI person should know the stages of grieving and loss as described by Dr. Elisabeth Kubler Ross: Denial, anger, bargaining, depression and acceptance. These stages do not happen in any predictable order. They flip back and forth. Emotional support and understanding should be given during this process. Grieving and expressions of anger need to be allowed. These stages may be different as individuals themselves are different. Life can ultimately get better, but not until the grieving is completed. To complicate matters, the patient's family and friends are going through this individual grieving process too.

Many health care professionals seem to forget about the emotional side of healing. They are at a loss and uncomfortable with emotional displays and often want to send you to a psychiatrist or therapist. Emotions are energy and need to be expressed so that steps toward a new life can be made. This is very important. A caring therapist who is knowledgeable about (M)TBI can be very helpful in this situation. Even a therapist who is not (M)TBI knowledgeable can be helpful The person with (M)TBI should be encouraged in their expression of feelings instead of being shunned away with demands of "Get rid of that anger" or "You'd better do something with that anger." Interestingly enough, it was only male physicians who ever said those things to me. Female health professionals I encountered always wanted to know where my anger was!

I did not recognize this new self attached to me, as it was continuously changing, but my old self was definitely gone. God was gone, too. I grieved for all these things belonging to my former "self." I wanted to protect what core essence of "self" was left, too. My life as it was before had shattered into pieces too small and distant to find. I still was not able to connect with many words to describe this to anyone. I still found myself terrified in the black abyss from time to time also. I walked and lived in a daze.

.

Brain Fatigue

When able, I would get on the computer to research information about TBI -- at least what I could comprehend. This activity really was a huge cognitive challenge, but nevertheless a place to start. If I didn't use my brain, improvement would never occur. "Use it or lose it,"as the saying goes. Even saddled with poor memory and comprehension, I collected TBI articles and tried to learn more about TBI and ways to heal. Despite memory and comprehension deficits, I hoped that things eventually would "click." I gave information to Dr. Pete whenever I discovered something important.

Dr. Pete's kindness was incredible, even to the point where he gave me his e-mail address so I could communicate between office visits. Writing and communicating was still difficult and I had forgotten how to spell some words. Proper grammar usage was a challenge. I would read and re-read what I wrote, in a fret over whether it made sense or was appropriate for the subject, hoping I wasn't saying or writing something rude or insulting. I could not judge those kinds of things anymore. The dilemma with reading, writing and not comprehending the impact of my words or others' words scared me. In my past nursing positions, I was accustomed to sitting with patients and completing 30 pages per patient of required information during the admission process, and then I would write up reports for their physicians. Writing for me had always been a "no-brainer." Not anymore. Comprehensive reading, writing and speaking was hard for me. I could not comprehend the whole story expressed through words. I couldn't see the literary forest for the trees. I could understand words separately, but organized together in a sentence made them incomprehensible. Nor could I remember separate words, sentences or concepts.

Concepts escaped my understanding whenever discussed. Much better for my cognitive abilities were very concrete directions and talk of that nature. With conceptual talk, my brain would just become overwhelmed and I could not absorb any of it. So, I kept working on concepts by exposure to them to build up tolerance. It took an awful lot of energy, and then I would have to rest in a dark room with no sounds or sights. Sometimes, I remained there over an hour before I could be outside again with light, voices and activity.

Let me tell you what my "fatigued brain" felt like. It felt like a plate of gelatin being violently shaken. My whole body shook at the same time internally and externally. When that feeling engulfed me, I knew I had to lie down fast, not talk, not listen, and not look at anything. Over a few years' time, my brain eventually stopped bottoming out.

What is it like, being a person with a TBI who is trying to communicate? For starters, you listen carefully as the speaker talks to you. You listen especially well when they ask a question. You may ask them to repeat it, if you are able. It feels as though their words are falling into dead space sometimes. No spontaneous associations come to mind to allow a response. Then, trying very carefully to remember and *consciously* hold that information in your mind, you have to separately think of what you want to say. Once you have that piece, holding those two separate things together and remembering them, you have to find the right words. Now, balancing all of that at the same time, you need to get those words out of your mouth. This takes a lot of time -- more than most people are willing to wait. It is also an exhausting, very conscious, mental juggling act for the TBI person to attempt. How incredible, the things we take for granted.

One of the worst things someone can do to a person with TBI as they attempt to answer a question is to cut them off by interrupting, or to abruptly end the conversation because it's assumed they have no answer. So wait. Wait at least three times the amount of time you would wait for anyone else. Also, refrain from quickly changing subjects, as his/her mind may go blank when you try to "change horses in midstream" so quickly. They may simply tune out and give up because it all requires too much energy to start the process over again. Exhaustion in communication attempts is common in recovery.

An extremely helpful maneuver is to have all other kinds of sound, lights and movement stopped when you are speaking to somebody with a TBI. No TV, background music or noise, and no hand gestures, either. I remember, about a year after my TBI, I was invited by a friend to lunch at a restaurant. Amidst her high, excited voice and her hand gestures, the clatter of dishes in the background, the waitress interrupting to get an order, people seated around us and music in the background, I started shaking because I was on overload. I couldn't focus on any one thing. I couldn't remember a thing. In response, I smiled a lot back then to cover up my weakness and confusion. No one could have any idea of what was happening inside of me.

At my chiropractor's visits, if he talked to me while working on me with his hands during an adjustment, I couldn't respond. I couldn't take in his words. Hearing his voice, being touched and hearing the paper underneath me crinkle simultaneously precluded either comprehension or communication.

Here's another example of what this is like: Think of the word, "Halloween." What automatically comes up in your mind? Autumn, kids, pumpkins, apples, the color orange, falling leaves, cool air, trick-or-treating, witches, goblins, black cats? For a person with TBI, there just are no automatic associations like that. When someone says a word, nothing lights up in your brain spontaneously. Absolutely nothing. A blank slate. So you search your mind, because you know it's there, but you can't find any information. You get scared because you possess an acute awareness of your deficits and you know it's not "normal."

If the person with TBI wasn't so aware of his or her own deficits it wouldn't be so bad. But many of us are acutely aware. It is terrifying to be so aware of your own mental and cognitive deficits -- all the more because you are unable to communicate this to make people understand.

My Changing Self

A concern towering high among the others was that I didn't know who my "self" was anymore. My old self was shattered, and everything familiar in my life had evaporated. I lost a connection to who I used to be. I knew who I used to be, but could not emotionally or intellectually connect to that person. I felt to be an entirely different person. I worked over 20 years as a nurse, but could not remember the things I used to be expert at. There was no longer a sense of even mastering household chores, and I wasn't positively sure of anything I did anymore. My memory was so bad that one time I went out for a walk and the thought occurred to me that maybe I forgot to get dressed. I looked down, relieved that I was!

Everything else remained in a state of disconnect, too. My perceptions were constantly changing. Things that I never would have thought acceptable before the TBI now made total sense to me. I understood things in ways I never knew possible. This was frightening as well as fascinating because every day seemed to bring a different perspective to some aspect of life. It was just like holding a diamond and looking through a different facet, through a different angle each day. That's how my perspective on things changed on a daily basis. I never knew what the next day was going to hold, but there surely was lots of introspective insight! This metamorphosis went on and on for at least a few years. After a while I actually became comfortable with it, able at last to relax and enjoy the changes.

My emotions remained very changeable. I learned that emotions and perspective are never written in stone. Never. We cannot count on them to keep us grounded. They are fleeting and ephemeral. They can be wiped out in an instant by being rear-ended in a car accident, having a stroke, or from suffering the explosive shock of a detonation device. Just as I had been so sure of certain things before my accident, I now learned that certainty had flown out the window. Life really is an illusion. I lived the shattering of the illusion I had called my life. I was always so certain I would always love my kids and husband, that I never would get a divorce, and that I never would let go of my faith. These were now in jeopardy. When your brain takes a severe hit, life can and does change, and your absolute certainty can all disappear through no fault of your own. Then you have to rely on will, because your emotions and perceptions cannot carry you through. You have to be strong-willed and determined to stick with something because you *knew* it to be right -- even if your new and very strong emotions and perspectives are telling you something drastically different now. For a long time, life after TBI is like walking on sand -- it shifts.

I learned to not judge. That was one big lesson learned from this. The other thing learned is that if you wait, things will change, your emotions will change, and your perspective will change. What you hold dearest in terms of values and priorities can all be taken away in a split-second. Just experience a car rear-ending yours, causing a TBI, and you will see for yourself.

Another advantage with this (M)TBI non-frame of mind is that you are always in the moment. You know how people are always saying, "Be in the moment?" Don't think of yesterday or look forward to tomorrow -- live in the NOW? Well, the only existence all TBI people have is the NOW. With no assurances about tomorrow, with no good working memory about yesterday, or even five minutes ago, the NOW is all you have. It is said that life as you live it is a gift. Believe it. It is said your faith is a gift, as well. Believe that, too. That can likewise be taken away with a (M)TBI. All of this adds to your fear, as you don't know what else may spontaneously be removed.

I was actually learning to enjoy this added, changing perception of things. At some point, I learned to just accept things and my new "self," as well. I could not waste any more of my vital energy on fighting. I needed to accept the situation so I could rest and conserve my energy for healing. I was curious if my brain and mind would ever settle. What would it settle on? Sometimes my new perceptions were disturbing to me and set off my emotions. I just wanted to be cognitively and emotionally stable. Stability was a long way off. I was involved in the rebuilding of my new personality and life.

Unpredictability became the norm. Some days I was caught up in some sort of daze and a dream-like state. At other times I would be down-to-earth. It was confusing and it was tiring. Nothing could be relied upon if "self" could not be relied upon. Therefore, all I could do was try to enjoy things, emotions, perceptions and people, for what they were in this moment, because it would all change again tomorrow.

On one occasion, my family took me to lunch for my birthday. I was lost in the sights, sounds and activity of people in the restaurant. I noticed a man looking at me from another table. Immediately, I thought I had done something to cause his stare. I wondered if my voice was too loud. Had I said something without my own knowledge? Did I do something inappropriate? Why was he looking at me? I must have done something. I asked my eldest daughter, had I had done something to cause him to be looking at me? She reassured me that I was fine and had not done anything.

Having someone to confirm and affirm your perceptions of situations is mandatory and very helpful. A reality check is good. It is especially helpful because you have to gain confidence in your own abilities and perceptions again once they do settle. You have a brain which is wired differently now. It works in such a different way that you have to get to know it and know also how to use it. It does not behave and work in the predictable way it previously did.

Most people possess an internal sense which says their actions and thoughts are ok and normal. That sense was totally absent for me. More than once during my recovery I would hear or think something and wonder -- did I say that? Did someone else say that? Did I imagine that? Did I dream that? Did I think that? How did I hear of that? How did I get that thought? My first recovery year progressed day-to-day like this. If I did or said something, I wasn't sure if it was "appropriate" and on some days I didn't remember at all. When I was tired, these concerns were more pronounced. Clarity grew much better as time went on and more healing took place.

You can research information on the brain in a good anatomy and physiology book, or go to the web at http://www.healing-arts.org/n-r-limbic.htm where you can read "The Limbic System: The Center of Emotions" (by Julio Rocha do Amaral, M.D. and Jorge Martins de Oliveira M.D., Ph.D.), and this changing self will be better-defined. You will see that there are different parts of the brain which govern and influence your behavior, emotions, perceptions, sexuality, and even your sense of self-preservation. For many people with (M)TBI these areas have taken a hit with ripped axons, neuronal death, edema and swelling from their injury. Consequently, their response to various situations might seem strange, "a little off," or exaggerated, but maybe not to the point of interpretation as nonsense, either, as the brain still functions. Certain areas of the brain (for example, the limbic system) recognize angry or fearful situations, assess danger and play a big role in our fearful responses. This area interacts and communicates with the other thinking and reasoning parts of the brain. If all functions adequately, a person can clearly assess and evaluate a problem, can separate their emotions from a situation, and can control their emotions, too.

Another area within the limbic system resolves our conflicts if we feel frustration. There are extensive connections between our limbic system and other parts of the brain, allowing each system to integrate and respond to different kinds of environmental stimulation. All of this interplays with blood pressure controls, heart rate and respirations controlled by the brain, too. Those centers are physically located close to the emotional centers. It's no wonder that many who live through PTSD or TBI experience high blood pressure when they never had that problem before. Sensations, sensuality, sexuality, conscious thoughts, hunger, how inhibited you are or are not -- all of these entities are governed by a complicated and great brain circuitry. The limbic system is sometimes called our "primitive brain" and sometimes this primitive brain might be overactive or underactive when we have a brain injury. Sometimes, other areas of the brain take over. Who knows for sure how and what takes over? It all depends upon which areas are injured. However, it is common that for most people with (M)TBI their behavior is sometimes a little or a lot "different," and sometimes the inhibitory area does not work the same compared with how they used to act before they were hurt.

Some people with brain injuries act out sexually/romantically. Usually, this is attributed to those affected brain centers governing sexuality and inhibition. We see this frequently while working with brain-injured people. I remember in my early nursing years how men who had been in car accidents or had brain injuries would always seem to be the ones who grabbed at the nurses, or even "fell in love" with them. Some people mistakenly think they exhibited these overtly flirtatious ways before they got hurt, but this is not always true -- and it is not restricted to men, either. I have watched men and women with TBI "fall in love" a lot. An influencing factor could be that a person with (M)TBI might be experiencing more highly acute sensations as a result of their injury. Then add to this the tragic fact that people have walked out of your life because they can't deal with the injured you (and this is common), and also you may be experiencing some real abuse at the hands of your insurance company, besides suffering depression and other problems. The stage is set for the time when someone comes along who shows a genuine, compassionate interest in you, the TBI sufferer. With the physical brain injury on top of all else, it's easy to develop romantic feelings for somebody who is nice to you. All of the above is my theory on why some TBI people become easily infatuated with others.

Well, living out these brain changes can be quite an adventure! I can laugh when I say this now. You really do gain a new appreciation for the intricacy and higher working levels of the healthy brain. It leaves you with a sense of wonder at how easily it can all change, too, and the ways it changes you and how you act.

This post injury recovery time, when all your cerebral functions are in flux, makes one feel like Alice in Wonderland. It's not all "bad" either -- it just means you are different. With your inhibitions knocked out, you may feel much younger, but with less restrictions. You may view situations in a whole new light. When you are able to think again, your view of life situations may be clearer than ever before, and you don't hesitate to speak or act on what you now perceive. Problems may occur, however, when your speech or actions conflict with those around you. But sometimes, if you are able to explain to others, you just may enlighten them via some new thoughts and perspectives, too. The brain-injured person is not entirely wrong! He or she just thinks and sees the world "outside of the box" now.

This brain injury gave me some great new adventures I never would have had otherwise. At the same time, I'm sure those who knew me were cringing or worried about how I was acting because I was not "right" anymore. Most everyone has heard the saying, "He's never been right since that car accident." I prefer to say, "He's never been the same since that car accident."

My brain's "inhibition center" did take a big hit with the brain injury -- but it does not mean that my intellect had flown off to other pastures, either. Not at all! Early in my recovery years, my daughter had fallen off her bike and broken her arm. My husband was taking her to appointments to get x-rayed and casted, the normal things anyone does with a fractured arm. One day, he came back and said the doctor reported a "bump" on her arm, but it would "grow out with time as she grew." I did not like the sound of what I was hearing. It sounded to me as if her bone was healing incorrectly, out of alignment, and the doctor was willing to let it go. This was my child and I did not want her care to be slipshod.

I phoned the doctor's office and demanded to speak with him. The doctor got on the phone and I told him I wanted her arm "healing the right way or all hell will break loose!" I NEVER talked that way to a doctor before. The doctor asked that my daughter and I come to his office immediately, so off we drove to the orthopedic office.

As if a red carpet had been rolled out for us the moment we stepped into that office, there was no wait time in the waiting room and the doctor was available immediately. I told him again that I was not willing to "let the bump grow itself out." My daughter's cast was then removed and her arm x-rayed and examined with me present. The bump appeared quite raised up underneath her skin and was boney. The x-ray disclosed how the bone edges had, in fact, healed crookedly. I stood behind the doctor as he examined the x-ray. I pointed out to him the place where the bone healed crookedly.

I asked that doctor what would he want if that was his child? I told him I wanted nothing less than that for mine! He closed the door of the exam room and stepped back. He asked me what kind of background did I have, that I knew these things? What kind of orthopedic work had I done, he asked me? Who are you? he asked. I responded that anyone with two eyes can see the obvious in that x-ray. I explained that I was taught the "old school" way -- you do things right the first time, and that's what I expected from him. I told him I had no particular orthopedic background, either.

I continued on a tirade about the insurance companies calling the shots now and doctors being dictated to by insurance companies, and the bottom line always being money -- with the patient being of secondary concern. He looked at the floor silently throughout my rant. When I stopped, he looked up and quietly told me that if this were his child he would want the same thing as I. He stated he would have a surgery date in a day or two, as soon as possible. He promised he would put her under general anesthesia and just lean gently on her arm to re-fracture, reset and cast it in a straight position in the OR.

And that is just what happened. Two days later, while my husband and I waited in the surgical waiting room, and my daughter under general anesthesia, the orthopedist came out, still wearing his scrubs. He showed me a new x-ray film and asked if her bone was straight enough for my satisfaction, because he was ready to cast it. This time my daughter's arm healed without a bump. The point here is that I never, ever would have mouthed off and demanded of a doctor in the manner I did after my brain injury! My inhibitions were totally gone.

Another incident occurred when my husband took me to Quebec one summer -- just as my French was coming back to me full-force. My depression had been terrible and he wanted to do anything to get me out of the dazed, depressed state of the TBI mind I exhibited. At the time, I really didn't care if we went or not, if I was alive or dead, that's how depressed I was. But we did go and it did help me immensely.

I had always been kind of a quiet, shy person who would never go up to anyone I didn't already know. However, in Quebec I approached total strangers, speaking in French to them and holding conversations! Totally out of character for the person I used to be. My husband just hung back and watched. I didn't see any of this as a problem. I was different now and there was no way I could connect back to who I used to be. I did not know where it was all going to lead, though. I felt acutely aware that maybe the people in my life might not like the "new me." A few of my friends, though, told me they think the new me is great. And those who never knew me prior to the injury don't see anything out of line at all -- so somehow it all worked out.

I could not communicate these "interior" changes about myself to anybody, nor did I have words for this experience. There was no person to look after these alterations for me or to direct me through this continually changing "self." I wrote my feelings in French in a journal every day. I wanted to protect my thoughts and what little "self" I had left, so I did that in a foreign language. The world, especially the insurance company, had been very brutal to me and for a long time. I thought if anyone wanted to strip away and trash what little was left of my innermost being, they would have to translate French to do it.

It was driving me a little crazy, not feeling any connection to who I used to be. I wanted to be someone, so I set about forming a new "self." I did this during walks. I took stock of what I could neither do nor have at the current time. That "wish list" included a good, reliable memory and an ability to function again as a nurse. I couldn't remember all the things I had been proficient in, even with an extensive background in hospice, home care, allergy and environmental medicine, nutrition, and also a knowledge of homeopathy. Truly, I reached the point where I said good-bye to my professional health care life and work.

But what could I be, I wondered? I thought I was good at smiling, and I did ok when trying to do small things for people. I was friendly. I used to be very independent, but I could no longer be that. Well, I could walk, thank God, I could still walk. So, I would walk and smile and say kind words when I could. I would do that and be good at that and not be angry for having to be dependent. I would continue to be outside, appreciating sunrises and sunsets, and plants, trees, birds and animals. That was it. That's what I settled on during that time. I told myself I was lucky to be in the NOW -- something that most people were always running around, drastically in search of. I was in the NOW and had been for a long time. I had no choice in the matter. I decided to enjoy and appreciate it. It felt good to have some kind of an outline for my new "self." I could be a good person with that, I thought. I would learn to rely on the goodness in others, just as I had to rely on the goodness within me.

Therapy

Dr. Abby, my new primary doctor, advised that it would be wise for me to consult a therapist, back when I was having problems with PTSD and panic. That sounded optimal to me, as any modality that didn't involve stuffing me with medications was worth a try. Subsequently, I went to see Mary, a social worker therapist who was absolutely wonderful. She gave me time to talk and did not interrupt. She patiently waited while I found words. She didn't push her agenda over mine. She helped me to grieve and let go of my "old self" and life. She helped me to take stock of what was left and asked what goals I set for myself. Mary taught me how to focus and ground myself to weather the panic attacks I was still occasionally having.

I told Mary my biggest problem was the insurance company. If they would just lay off, I surmised, I could let my guard down enough to heal. They were robbing the energy I needed to get better and replacing it with anxiety. Defending myself continuously had weakened me severely. No one really understood I was literally fighting to survive.

As it turned out, Mary also encountered the insurance company at their best. They told her I was "adverse." Apparently, trying to defend yourself so you can rest and recover from trauma is labeled "adverse" by those in charge of medical claims. I told Mary I felt like road kill with them picking at my remains. After taking a lot of time to listen, Mary said to me, "You would have no problems with them at all, JoAnn, if you would just allow yourself to be dehumanized." We both laughed at the absurdity required to deal with them. Mary was wonderful and understood how I was protecting my "core self" and the dignity that this insurance company was trying to destroy. I shared with Mary what I had left of a "self," and she helped me to expand upon it.

She instructed me about the good power of my anger, its value and how to use that power for healing while grounding myself to ride out the waves of panic. She showed me how to get a little bit more control back in my life. I learned how to endure a firestorm, and in the process transform "self" when nothing else could be done. Like an alchemist, she showed me how to sit tight in the fire and how to decide for myself which transformation experiences to accept. Mary and I worked something akin to magic together with these therapy sessions. She explored with me all those emotions that were mixed up and tangled together, teaching me strategies to deal with them when they reared up unexpectedly. Ultimately, I gained strength enough to voluntarily stand at the edge of that bottomless black abyss, at last prepared to look into the darkness at whatever was there that scared me so much. Mary, like Pete and Abby, is one of those angels who touched my life. I had said that God wasn't listening to my prayers initially, but He did come through loud and clear with those individuals.

None of the three was paid by the insurance company for a very long time. I was forced to go with my attorney to a Workers Comp hearing to find out why, and this conjured up a ton of fear for me and consumed incredible amounts of my energy. Any sort of communication with that insurance company, even just receiving their letters, played on my severely damaged nervous system and PTSD. I didn't think I would ever be able to look at their company name without shaking. Now, I had to confront them in a Workers Comp hearing room with a judge. It was an overwhelming experience, but I got through it. Of course, they had to pay, and they did -- but not until they were forced. Oh yes, what they force you to go through to get the help you need! They bank on wearing you down and giving up so they don't have to pay out benefits or arrange medical help. It's no wonder the suicide rates are high when someone is badly injured, unable to work and forced to deal with Workers Comp. I will never forget Mary, Abby or Dr. Pete for all the times they went without pay and told me not to worry about it, that we would just focus on getting me better.

Dizziness and Memory

As recovery slowly continued, dizziness gripped me severely. It began one day in the early hours of the morning, when I woke up in distress. When I opened my eyes the room was whipping and spinning around like some amusement park ride gone wild. I grabbed at the sides of the bed and held on tight, it was that intense. I hadn't even sat up yet. What was my brain doing? I immediately feared a seizure or something similar to that. Later, I went to my chiropractor and asked him if there was anything he could do for it.

Dr. Pete explained the dizziness was a residual from the brain injury and that we could do rocker board therapy. He worked with me on "short foot" exercises. Short foot is a way to stand on the heels and balls of your feet. When that was mastered, I would stand in that position and close my eyes while someone lightly touched me in different places on my body. What a workout for my brain! I became dizzy from a few minutes of that and had to rest until the room stopped spinning. The exercises involve the proprioceptors on different parts of your body and how they tell the brain where your body is located in space. There is an incredible feedback system between your brain and nerve endings on different parts of the body. When this system gets damaged, as in a head or neck injury, you can experience a lot of dizziness.

Once accustomed to standing in that position, feeling the light touches and not getting dizzy, I next stood on the rocker board and had to hold my balance. A rocker board is a flat wooden surface on which to stand, which is mounted on top of another piece of wood which happens to be convex – like a half a sphere. The rocker board rocks. You stand on top of this in short foot position with knees slightly bent and hold your balance. This was another cerebral workout also productive of dizziness. I gradually increased my time as I could tolerate it. Then I added complexity to standing on the rocker board and maintaining balance. One of my kids would lightly touch me on my shoulders, arms or legs, head, etc. I would have to hold my balance while getting that kind of stimulation. Over time, I would add more advanced forms of stimulation. I would bounce a ball or throw one in the air, catch it and still hold my balance. Another tactic was to put on music and sing while holding my balance on the rocker board. Another proprioceptor exercise was to sit on a chair and roll the soles of my feet upon various sizes and textures of balls. All of these things stimulate the brain and strengthen this proprioceptor/brain connection. All of this helped in taming those horrific dizzy spells. Improvement came within weeks and the dizziness actually went away.

A few years later dizziness did return, however not as severe as the first time. Dr. Pete did a Dix Hallpike (Nylen Barany) test for benign paroxysmal vertigo with me at his office and determined otoliths might be the culprit. Otoliths are small fragments of bone free-floating in the inner ear which disturb the equilibrium mechanism. This is common after head and neck injuries from auto accidents. Dr. Pete did Dix-Hallpike maneuvers with me and taught me how to do them at home. I have not been plagued with dizziness since.

Losing a reliable memory is terrifying. Mine was not in good shape at all. I happened upon a book called **Depression, Cured at Last!**, by Dr. Sherry Rogers, M.D. I had read many of Dr. Rogers' books in the past and now she wrote a book about the brain! I skimmed though the book

Brain Jolt

because I could not handle the total comprehension involved in reading it or remembering all of what I read. I was especially interested in the pathology of what she was writing about, and I always loved reading pathology anyway.

To my surprise, a portion of the book discussed the role of phosphatidyl choline in the healthy and stabilized cell membrane. A realization struck me that it would only make sense that my brain cells needed stabilization after all the physical trauma I had endured. Phos chol (short for phosphatidyl choline), which may be better known to you as pure lecithin, is very important. Generous supplies are needed by the brain, as it is the substance from which acetylcholine is made. Acetylcholine is responsible for much of our brain chemistry and it is a major brain neurotransmitter. There is no way I can explain for you here all that is written in Dr. Rogers' book. I encourage you to read it for yourself to understand this information and more. I decided I would buy some phos chol and use it. I certainly couldn't lose anything in trying. Conventional medicine had no answers for curing brain trauma or improving memory.

The company producing the purest form of phosphatidyl choline (that I know of) is Nutra Sal. I bought a bottle -- relatively expensive at about $80.00 per eight oz. My brain was definitely worth it, especially if it worked for me. I took it according to the directions on the label. Within four days I was incredibly more focused! I was so happy at the difference a food supplement made! I called it my "brain glue." The difference was like night and day. I encourage you to read and research phosphatidyl choline yourself. It is important to know that this substance is found in our normal foods but, unfortunately, they are foods that most people don't eat much

83

of anymore. Cauliflower, liver, beans, peanuts and egg yolks are the ones rich in lecithin that I know of.

I did not have to stay on phos chol forever, either. I stayed on it to achieve a good, stable, focused state of mind, and then tried to slowly decrease the amount I was taking. It took many months, but after a time I was able to halve the dose and maintain a good brain function -- then I decreased it some more and maintained, too. Eventually, I got off of it. But it is a food supplement offering good nutrition, not a pharmaceutical drug. Therefore, there is no insurance compensation for this. You pay out of pocket. But I have found many vital, healing substances not covered by insurance companies.

Good nutrition is absolutely essential for good brain function: Organic, if possible, whole grains, fresh vegetables, organic greens, fruits, beans, nuts -- and organic meats, if you are not vegetarian. A good clean water source is necessary, too. Throw out the processed and junk foods -- and I am serious about that -- if you really want to heal. I used supplements, as I had always used vitamin and mineral supplements. I also used homeopathic remedies to help me heal quicker, too. Homeopathy is not something recognized by medical science to be valid, and depending on which state you live in, it is not likely to be covered by your medical insurance, either. These things that I did were done on my own, without a physician directing me or giving me some seal of approval on each. I take total responsibility for my own health.

So, the phosphatidyl choline was a godsend, as were homeopathic remedies. The remedies were of great help. Notice I say they were of help, not that they were magic and returned me back to who I used to be. They did not "cure" me, either, not at my deepest levels. I also had a friend who did craniosacral massage. Her sessions helped me feel so much more grounded and my jangled nerves could de-stress and relax with that.

There was one other thing I did for myself to get my memory back. When my ability to sleep a little longer returned, my dreams eventually came back (what a gift dreams are!) and they were all in French. I used to be a college French major and had not spoken, read or listened to that language since 1976. But my dreams now were all in French. It all just flooded my mind and felt so soothing to me. I began speaking French at home. My husband went out and bought me French music. I balanced on the rocker board and sang in French. I would go into Dr. Pete's office and speak French. I always loved that language. I decided I would use it to help get my memory back.

I obtained the lyrics for a French song and wrote it on paper. I took one sentence and wrote it 10 times, just like in elementary school if you got something wrong on a test. I copied it once on a piece of paper and carried it in my pocket whenever I went for a walk. I repeated out loud the first three words, and then repeated them silently many times. I stopped and walked a distance and then tried to remember those three words. If I could not, I would retrieve my paper, read and re-read and repeat and repeat. Eventually, when I spoke the first three words, I would add another few words. This was a very long process and very discouraging, yet I was determined to remember. I tried to remember those same words the next day, and many times discovered I had totally forgotten about them. So, I would go at writing, reading and memorizing the same words again for yet another day. I thanked God that I had nuns for teachers in my early years who were very strict in elementary school. The lessons they gave us in studying and disciplining ourselves to do our work and memorizations were invaluable. Here I was, 40-something years later doing it all over again on my own. I had no therapist to direct, and I had no one teaching me to do these things. I was drawing on prior life experience and what I knew to be right. Back in the 1980s I worked with a physician who had always encouraged me to go with my gut intuition on many things. Gut intuition was what I was relying on now, since my memory did not work. Some friends seemed very upset that I wasn't doing things in an "approved medical" manner given by a specialist physician, but I knew within myself what I was doing was the best way to go. Besides, what do you do when conventional medicine has no answers and even fails to recognize problems? What do you do when the specialists you would like to see are not sanctioned by an insurance company and you are stuck in a system? Now, seven years after the car accident, I still make lists to be certain I do not forget things, but all in all I would say my memory is on a level with my friends, depending on how tired I get.

Keeping lists and keeping calendars is very important when you are recovering from or living with TBI. In the very beginning, though, do not be surprised at all if the TBI person you are living with does not even remember their calendar or list, even though they may have helped you plan it. I started off like that, as do many. It takes a very long time for all things to fall back into their proper places and reorganize in your life. Patience is the biggest thing with TBI. You need many golden years of patience.

Healing With Spirit - The American Indian

"...Oh boy, you've no idea what I've been through

Oh Lord, I feel so stuck that I can't get to you..."

-- Lyrics from George Harrison's song: "Looking for My Life"

The American Indian didn't want his name used in my book. It was something he made clear when I told him I might be writing one. He thought the book a good idea, as it would help other people. His reason for anonymity had nothing to do with shyness or fear of discovery. He insisted upon anonymity because in his tradition healing is a gift from The Creator, and credit and gratitude should be given to The Creator.

It was February of 2001, and my journey of healing had already taken a long road since January 15, 2000, but I had a very long way yet to go. I continued the struggle with my memory, insomnia, cognitive skills, depression, feelings of betrayal, different emotional rushes, and a total disconnect emotionally from others and from God. I had a lot of anger with God. I did not trust Him. My brain, neck and back still hurt a lot, though improved. I still took no medications for the pain. These were the days before phos chol, and I still sat around with a blank mind and had a hard time planning and then carrying out any kind of project, no matter how minor it seemed. I had nothing to look forward to. It seemed the only thing I could do well was walk and appreciate being outside. I continued cervical traction at home and used the rocker board for dizziness but I still existed in a daze, and state of disconnect. Interestingly enough, when I went for walks, a mental picture of a phoenix rising from the ashes would spring up. I thought that image to be myself, but I knew I hadn't totally risen from the ashes yet.

I was not well enough to work and I was not happy being dependent. I needed something to do that would stimulate my brain cells to work in a meaningful way. Because I had been studying homeopathy for many years prior to the accident, I decided I would go back into that. I needed some kind of program allowing me to progress at my own pace. I could not tolerate riding any great distance in a car, even as a passenger, for all those minor bumps in the road tormented vertebrae compressing the nerves in my neck. Pain would still go into my back and down into my arms. Occasionally, I would lose feeling in my arms, hands, fingers, legs, feet and toes. I still went to Dr. Pete regularly for adjustments and he always pulled me out of trouble caused by the paresthesias and pain. I was not yet cognitively sharp enough to sit in a classroom and listen for any length of time to a lecturer and be capable of a prolonged, detailed conversation with anyone. I was still overloaded with stimulation and experiencing "brain fatigue." I still was not sleeping all night, either. Whatever I decided to do, I had to start very slowly. I signed up for a distance learning course in homeopathy with a well known homeopathic school, whose credentials always impressed me. Concurrently, I received help in the form of remedies from a friend homeopath named Michael from a different city in NY State. The remedies were helping. Unfortunately, because of severe winter weather and driving conditions, I could not visit him for evaluations. I was stuck without a homeopath, it seemed.

After registering with the school, I met many international homeopaths, and a good many also lived in the U.S. A few privately offered to help when they realized I was recovering from a car accident. I decided to accept help from an American Indian man whom I had befriended and who also knew homeopathy.

I sent him my case and the history behind all of it. He replied that he prayed for help before he worked on any case, and asked me if it would be ok for him to do that for mine. I was very angry at God, and thought it a great possibility that He did not exist in the manner I had been taught, or else He would not have let this nightmare happen to me. When the Indian told me he needed to pray, I said that was fine, but I thought to myself that if that's what he needed in order to help me -- well, so be it. I didn't care one way or the other if he prayed or not.

That very night, I awoke every hour on the hour. I could hear -- I heard my Indian friend's voice calling my name periodically throughout the night. It felt crazy. He lived in a different state. I thought it was the strangest thing. I thought my brain was really doing wacky things and I laughed it off, wondering what else my brain was going to do.

The next morning was Sunday. As was my habit, I went to visit my 91-year-old grandmother to have tea with her. She sat down at the kitchen table and looked at me with very wide eyes. She said, "Something very strange happened in the early hours this morning. I had a dream, but it was not a dream because I was not asleep. I heard my mother calling my name (her mother died in 1969). I saw her, in my mind, she was outside and she had you standing in front of her with her arms crossed over you like she was protecting you. She was dressed the same way as if she had been to church. You were just a little girl standing there in front of her when I saw her early this morning. You were the only one she used to take to church with her when she was alive. She kept calling, 'Wanda, open the door, its cold out here, Wanda, open the door!'

"So, I went to the door because I heard her and because what I saw in my mind was so real. I went to the door to open it, and I looked out, but there was no one out there." Grandmother looked at me with a very shocked expression. Her eyes were as big as saucers. She said never in her life had she experienced anything like that before. She asked me, what did I think of that? I told her I had someone who was trying to help me recover from this accident and he told me he was going to pray for me. I did not tell her I heard his voice calling me practically every hour all night, too, while she had her vision.

When I got home, I contacted my Indian friend and asked him who he was. I told him I thought his praying woke up my great-grandmother who had been dead since 1969. I told him my grandmother saw her mother and I together. I told him I heard his voice calling me throughout the night. I received a long explanation that he was a spiritual leader of his tribe, that he was trained by his father who was trained by his father, and on and on his traditional healing lineage went. I had never known an Indian healer before. I was amazed.

We communicated every day with each other, sometimes a few times a day, especially during that first year. My American Indian friend visited me in many dreams over the following years. He watched me and protected me in them. I could hear his voice many times internally during the day, as well as during the night. Sometimes I could feel his presence. It was at those times he told me he was praying for me. I felt safe with him. I could feel myself changing inside more and more as I worked with him. My anger was lessening, my tension loosening, and I started to trust again. He even restored the dignity I lost through the degradation endured at the hands of the Workers Comp process and the insurance company. We worked together for years. A deep bond developed between us. We called each other brother and sister.

It was not all one-sided, this work that we accomplished. I had to work on my own, too. He gave me a meditation and prayers to do daily. As hard as it was, because I couldn't feel a sense of connection as my cognition, comprehension and memory were so poor, I still did as he instructed. In the past, before the car accident, I meditated daily, sometimes twice a day. Not being able to do that again was another monumental loss in my life originating with this injury. But I remembered what an old friend, a nun, told me many years ago. That advice was, "Act as if, and you will become." So that's what I did. Even though I saw no direct results, more than half the time sitting with an absolutely blank mind, I kept on with it. With the prayer, many times I was still angry at God, but I kept on. I acted as if – trusting, somehow that it would all work out – so that I would "become." This anger with God and with others was a huge obstacle in my life now. My Indian brother prayed for me and helped me get reconnected. It was a process. It did not happen immediately and it did not happen quickly. He had lots of patience. He worked on my physical pain as well as my emotional pain. He gave all credit to The Great Spirit.

He sent me e-mails of Joyce Sequichie Hifler's "Cherokee Feast of Days." He taught me how to view the world in a different way, according to his tradition. I was badly in need of a new view of the world because my world had shattered. His world was so much gentler. Everything was connected in his world. Everything had a purpose. There was no good or bad. Things just were. He taught me to suspend judgment on things. He told me to accept the shadow side along with the light because they belong together. You cannot have one without the other. Don't judge as "good or bad." He explained to me how we are all connected, and everything that is alive is connected. He knew of my feelings of betrayal and how I was left alone when I needed someone when I was injured. He taught me how I am never alone. He showed me how to reach into my heart and feel the connection with Spirit and with all those who live in Spirit.

I learned that good medicine is the blue sky, the wind, the plants, the trees, the birds in the sky and all that are living. I could connect to those whom I love in my heart. Everything on this earth and in the spirit world is one in Spirit because it all comes from Spirit. No one is ever alone. He taught me that there is no death for that very same reason.

I dreamt many dreams with my Indian brother in them. One of the most poignant was when he found me in that bottomless, black abyss I was so terrified of in my early recovery stages. He was in there with me, pulling and pulling at me to get me unstuck. Another time, he stood before me in a dream with his hands out, coaching me and encouraging me to take more steps toward him even though I was scared half to death. I have had many dreams with my brother in them and each is a treasure. We went on for years like this, working together, he and I. I was challenged in many ways to become not only physically better, but better as a human being.

My neck and brain healed, as well as my spirit. I learned to go into my heart to discern what I needed to do in my life. I learned to better control my runaway emotions, rather than allow them to control me. He taught me that emotions are not to be suppressed, but expressed and let go of. He taught me to be the person I wanted to be, instead of being tossed about by the changing emotions and circumstances I experienced. He helped me to function again with a new brain and personality. He helped me to restore good relations with my family and those around me. Every time I pray now, I give thanks to The Creator for my Indian brother.

I wish I could give everyone the same experiences I had with him. All I can do now is remember how I prayed to God for help in the early days of my injury. I was terrified of living and wanted to die. I called for help and none responded. Things were so bad that I thought either no God existed or the one who did had turned his back on me. Well, so I thought. I did not realize help was coming -- and that God had kept the best for last. He kept this one for the deepest healing possible. Looking back now, I see that I was not abandoned by God. My connection to God existed then, though I could not feel it. It exists now on a deeply felt level. I maintain a whole different view of whom and what God is. Many lessons were learned and not all of them can or should be put into words. There are some experiences and lessons in life that defy words.

I am reconnected emotionally to my own family, too. I hold no more anger toward anyone, either. There is no more fear, there is no more panic and there is no more PTSD. Seven -- lucky seven, some might say -- years have elapsed since my injuries, and I can say concerning all those who helped me through this ordeal, each one is a gift from God. I hold them all in my heart. The lessons learned from my brother during this honored and privileged time I have kept safe and close. I continue to use this gift for myself and for others, too.

*** *Ga li e li ga – I am grateful* ***

Back To Work

The healing road of a (M)TBI is very long. I guess in some ways it is best not to dwell too much on the time element. Sometimes it is best to understand that all of life changes in very unpredictable ways, whether you have a brain injury or not. I think what is most disturbing is that a TBI changes you and your life in abrupt ways that you or others never could have predicted or prepared for. Sometimes, knowing that recovery is a process and takes many years can make it seem too impossible, especially when you don't know what you will have or what you will be when you come through it all. You have to keep an open mind, however, that not all of these changes are "bad" and that time is on your side. Keep in mind what my Indian brother taught: "Do not judge good or bad – things just are what they are. Look for the blessings. Go into your heart and see what's there. Who do you want to be in this situation? Be that."

Existing for over a year in a daze with altered perceptions and no real working memory, flooded with constantly changing emotions, the thought of trying to find work "out there" is a huge hurdle. It may seem too difficult that such a thing as working can even be done. How do you get back into the working world? Who do you tell your diagnosis to or don't you?

For approximately one month before my car was hit, I had been having a continually recurring dream. It was that I was back working with a physician (Dr. Sally) whom I had worked with 20 years ago. She also had been my personal family physician since the 1970s. She not only helped me with my many allergies throughout my teen years and beyond, but brought me into the medical working world once I became an R.N. I worked in her office for approximately four years, straight out of nursing school. After that I moved on to work in other areas including hospice, obstetrics and gynecology and community health. It had been about thirteen years since I last saw either her or anyone on her staff, as I had also been using some other alternative health ways to maintain my own health. So, I didn't know what to think when I was having recurring dreams about working with her in her office.

Then the shattering event of January 15, 2000 happened. The injuries I sustained left me sleepless as well as dreamless for approximately a year and a half. How strange it was that my first dream was not only in French, but it was the same recurring dream of working with Dr. Sally that I had prior to my car accident! (I had been a French major in college, but had not spoken, read or heard the French language in 28 years! What a pleasant surprise that I was dreaming fluently in that language. How peculiar to be having the same recurring dream as prior to the collision.)

Well, the same dream persisted. I kept on having it night after night, and I thought I had to do something that might make it stop. It really was bothering me. One day after a chiropractic adjustment, I walked into Dr. Sally's office, thinking that I should see what was there. Maybe this dream would stop and leave me alone if I went into her office.

The "mud room"/entrance still looked the same and the emotions I felt seeing that familiar wallpaper and being in that building again felt good. I opened the door and walked into the reception area, not knowing what I would find. Bev came running toward me, threw her arms around me and exclaimed, "Are you back here to work with us again?!" What a shocker! Bev had been with Dr. Sally since the first day the doctor opened her practice back in the early 1970s. I was almost speechless to see Bev and I was flabbergasted to hear what she had said. I wanted and needed to do some kind of work, but I didn't know what I would be able to cognitively handle. I told Bev, "I have a brain injury." She burst out laughing and jokingly said that I would fit in well. After the laughter died down, I explained that I really did have a brain injury and didn't know if I could handle that intense and very responsible kind of work anymore.

Now, on that particular day there were no office hours as the doctor was out of town. I asked if I could walk around and see the place again. Bev said, "Sure – go ahead." Memories of many years came flooding back – from the earliest times when I was a patient there to when I had worked there professionally. That place just felt so good. I met Donna for the first time then and her words and smile were encouraging as well.

Bev kept after me about coming back to work and explained Sally's schedule. Sally had scaled back her practice and was seeing patients part-time. Therefore, they needed a part-time nurse. Again, I told Bev that I didn't know if I could handle it. I told her my brain had taken a big hit with that car accident. Bev and Donna both told me they would help me and not to worry – to just come back and try. I told her I would call her back later after considering it. Ultimately, I decided if the doctor wanted me back, knowing I had a brain injury, I would give it a try.

Back to work I went. The atmosphere of that office felt very good to me. These were the same people I had known since I was 16 years old, who cared for me when I was very ill, and had watched me grow up and start a family of my own. It just seemed very appropriate to be there again. They were like family to me. Sally was very welcoming, as were Bev, Donna and Sue. I stood there at the nurse's station looking in at Sally, who sat at her desk. She remarked that having me there again brought her back to her earliest years when she first started her medical practice. It was a sense of a circle being completed for me (and she, too). I was one of her very first patients when she started her practice in the early 70s. She not only was an excellent doctor to me and my family when I was much younger, but it was her office that I started out my nursing career in, also. So, it was an incredibly wonderful, strange sense to be standing there once again -- I felt blessed to be there. Another great thing was that many of her patients had known me from working there back in the 1980s. It was great reconnecting with them and getting caught up on their lives.

Something that did not resemble "old times" was my necessity in the beginning to consult Bev and Donna often for "reality checks" and confirmation of what I was reading in terms of doctor orders, how I interpreted them, and how I would go about my work. I was always right, however, in the beginning I needed to hear that I was right before I acted. They were wonderful supports. As time went on, my confidence grew and I was on my own, taking care of patients again as I had done many years prior, as well as doing the nurse consults, teaching patients, allergy testing, and working with the doctor.

When Dr. Sally did not need me, I was working with my nurse colleague and friend Lana, whom I had worked with periodically since my hospice years. After hospice, we both taught home health aide courses for the Red Cross, among other nursing positions. After accepting work at the doctor's office, Lana called to ask if I might work with her again, knowing nothing of the car accident. My response was, I did not know what abilities I could offer anymore. She worked at a home care agency and now she was asking if I would help her teach students and take care of patients again. I did return, on a part-time basis, on the weeks I was not working with Sally. Now, I had two nursing positions to test my healing.

It was a challenge in the beginning, dealing with the lights, sounds, and the many student personalities that were in that room all together, but I did well enough to stay on. The one thing I can honestly say is that both Lana and Sally only accept top-notch work, and so it was great for me to see that I still met the highest standards. That was very reassuring to me.

At the end of each working day, I really needed to rest. Days off were very necessary to balance all that brain activity. Going outside to walk, being in silence and just watching birds and seeing trees, grass and plants, as well as feeling the wind were all very necessary to balance myself. When I got home from work, I could do nothing else for the rest of the day. My husband and kids were great with getting meals together and pitching in with household chores. Things were coming back together slowly.

In the meantime, I kept on talking with my native brother and expanded on ways to view self, the world, this life, and the importance of a strong connection to Spirit and to one's relations. I persevered with the study of homeopathy and completed the diploma course and also the postgraduate course in homeopathy. My thesis was on healing traumatic brain injury with homeopathy. Beyond that, I revised two of their courses and am tutoring their pathology, anatomy and physiology, and diploma and postgraduate homeopathic courses, also. Yes, I am different and my brain works fine and is very reliable. It is important to feel that you can rely on your brain.

My relationship with my husband and two daughters continues to change and evolve, as all relationships must under all circumstances. My eldest said it best one day, "Yes, you are different, and we just had to get used to you."

We are very, very lucky as a family. Many families and marriages do not survive (M)TBI. We almost didn't, either. It really was the roughest thing we've ever been through. Many times, I thought our family was going to disintegrate from all the stress and haphazard winds of change. I have given thanks many times to Dr. Pete and my Indian brother for helping to hold me together during those critical years when I had literally wished to be dead. I am thankful to all who helped me during my long recovery process.

Now, I spend two days a week working as a nurse supervisor. An additional day is spent doing community teaching about traumatic brain injuries. I work primarily with traumatic brain injury clients who come to our structured day program. I can relate to where these people are and I try to use what I have lived to help them. My cognition and memory are just fine. I love my work there and I also love working with my online homeo-pathic students.

Family life is work, too. Reconnecting with my children and hus-band took time. Even though we lived in the same house all those years, this injury took an enormous toll on "us." I have focused on getting back to being a mom and wife.

How To Help

I intended this journey through my life's most tested moments as a beacon of encouragement and hope to those who endure a (M)TBI and to people who live with them. I want to enlighten those who have never experienced a (M)TBI because there is absolutely no way you can ever truly understand the brain's alternative world unless you have visited that bizarre place yourself. I hope my words and experience can help health care professionals to become more sensitive, kinder and less hurried when dealing with someone who acts a little "slow" with their words, or a little "confused" when suspected head and neck trauma are involved in their history. How much more humane it could be if employees in the insurance, medical, government, and legal systems would open their hearts and show some compassion. Instead, the normal procedure today is to deny medical help, benefits, and services, and automatically assume suspicion, fraud and deceit. These ignorant actions only wound more and prevent healing from taking place.

Unfortunately, this injury does not always show on a CAT scan -- and sometimes not on an MRI, either. The only people who receive this diagnosis without challenge are those in comas. No insurance company can argue with a coma! But if you have anything less than coma status, beware -- your (M)TBI diagnosis is likely to be disputed and accusations of fraud might be aimed at you. Or you may be given a different diagnosis – possibly a psychiatric diagnosis. (A physician I once worked with used to say that the medical profession gives a psychiatric label to anything not understood.) The gold-standard "scientific approach" to medicine is wonderful except for the fact that we don't know all there is to know about the body, trauma, or even about science. Our medical tests are not sensitive enough to show all problems either. Unless a medical diagnosis is backed up by a test, most likely the diagnosis will be disputed even if the person exhibits all the signs and symptoms of that diagnosis. I wonder how many people out there with (M)TBI have been treated as psychiatric cases when the problem is that they have sustained brain trauma.

I fervently wish for people to realize not only the extreme difficulties and duress that this injury brings, but also the obstacles caused by improper diagnosis and of the additional trauma and duress suffered at the hands of the insurance companies and other systems. Everybody deserves knowledge of the hell a family suffers simply because there are no supports in place for either the kids or the marriage itself. Indeed, there isn't even any family education provided by doctors about what to expect from your injured loved one, even if you are lucky enough to find a doctor who can diagnose correctly. Even knowledgeable physicians will say they cannot predict how any case will turn out.

An appointment with the neuropsychologist was not in the cards for me as a patient because of the insurance company's refusal to guarantee payment. But I did get to meet a neuropsychologist later on. My nurse colleague friend and I eventually visited a local hospital's brain injury unit to gain information about setting up a structured day program. The nurse manager there was very personable and talked with us in depth about how we might be able to help a few outpatient families there. When this nurse discovered that I, too, had recovered from a brain injury he was astounded. He asked me how I recovered so well and I recounted my story. He, in turn, related it to the staff neuropsychologist (most likely the one I would have seen, if not for the insurance company's deliberate and repetitive tendancy to ignore a written request confirming the doctor's payment). Later in the week, I received an invitation to talk with this doctor.

That day at the hospital was interesting because we did not know what each other looked like. There was one other female person in the waiting room when I entered. A man with the appearance of a professional walked into the waiting room and called my name out loud. He immediately walked over to the woman who sat there, hunched over with her cane. He repeated my name again to her. Realizing that this might be the neuropsychologist whom I had the appointment with, I stood up and announced, "Here I am -- I'm over here!" He looked over at me and his mouth dropped. Clearly, I was not in the shape and condition he anticipated!

The neuropsychologist and I talked for close to an hour and a half in his office. He wanted to know the details of my accident, how I was diagnosed, and the nature of my symptoms previously and currently. I described in detail all the trouble I had in acquiring a diagnosis, how my own physician had mishandled my case and how I absolutely refused to go back to him -- besides removing all our family records from that practice. I relayed how I was thrown out of an orthopedic surgeon's office when I refused surgery and how I relied totally on my chiropractor for everything until I was able to find a doctor who actually could put a name to what was happening with me. He shook his head and agreed that most physicians do not know how to recognize and diagnose a (M)TBI, and this was a huge problem. He told me that my story, unfortunately, was not unusual.

He asked about specific symptoms and I related the horrific dizziness, and I described what that was like and how terrifying it was when my brain "bottomed out." He stated that my descriptions were so on-target that he couldn't get over it. He assured me that only someone who has lived this could describe it so accurately, and informed me of the medical word for my brain bottoming out: "Brain fatigue." I told him how difficult it was to read, understand or find words to speak, and how my memory was affected. He nodded at everything I said. When he inquired what medications I was on and I said "none," he was shocked. He asked what had I used for the dizziness and I replied "rocker board therapy and Dix-Hallpike maneuvers" from my chiropractor. He had never heard of those, so I explained both. To satisfy his curiosity about regaining my memory in good working condition so that I was able to work as a professional nurse again, I informed him about phosphatidyl choline and homeopathic remedies.

I described for him how I would write sentences 10 or more times, or very short lists of words to work on my own memory, since I was allowed no neuropsychologist to direct me in these things. Asked about sleep habits, I described that in the beginning it was practically impossible to sleep for a long time and that I experienced no dreams, either, for a very long time. So, I just rested and slept when I was tired. Sleep was considered good if it lasted for two hours at a stretch. I was much beyond those days now, however.

He learned from me the horror of feeling as if my body was shot through with adrenalin from time to time. The fear and shaking was terrifying. Asked about my emotions, I told him how extremely harsh and uncontrolled they were and how settling those down required a long time, besides having to consciously work with self-control. I described the strangeness of having a different personality and inhabiting a different self, and how that self had continuously changed from day to day, and for the longest times, so that I could never be sure who I was. I told him at times I thought I was really going crazy, and he assured me that this feeling was very common, but that I was not crazy. He confirmed how often brain-injured people are treated as psychiatric cases or mentally retarded, though they are neither.

The physician asked who assisted me with the homeopathic remedies. I told him at first there was a homeopath in a different city, but later I went to someone else who helped me with healing. I told him, not quite sure of his reaction, about the American Indian man who used his traditional native ways to heal me. I explained that this man was a spiritual leader and healer of his tribe, as well as what might be called a "shaman." I told him about receiving Reiki, a Japanese healing technique, and talked about a friend who helped me with craniosacral therapy. The doctor said he was absolutely amazed. Asked if I was able to drive a car any great distance, my response was that I had driven my youngest daughter to Virginia for an extended weekend to visit our friends, and then I drove us back. He could not get over it, he exclaimed.

The doctor then explained what I "missed out on" by not seeing him or being treated by a "specialist" in TBI. The medications were all aimed at symptom control and did nothing to heal the brain injury itself. They use Ritalin and similar drugs to "push" a person's brain to counter the brain fatigue, so a person could function longer without having to rest. He told me antidepressants were prescribed for the depression and a sleeping pill prescribed for sleep. He informed me about the anti-anxiety meds implemented to help decrease the fear and other emotions. I was told about the narcotics used for pain relief of the same brain pain that I described having.

When I replied I was glad I didn't get that appointment with him after all, because I just didn't want all those drugs, he just smiled. When asked what was wrong with resting an injured brain instead of drugging it, he said nothing is wrong at all with that approach. Insurance companies, he added, want to see people functioning quickly again, so drugs are popularly used to "push the brain." But what can those drugs do to a ripped, bruised brain, I asked? He had no idea -- no studies have been done or released yet.

This doctor asked me what, in my opinion, was the toughest thing to deal with and the hardest to heal? I replied that getting over my anger at God was the hardest and the toughest. I told him how the Indian man helped me with that, and I could truly say that my anger was gone and I was at peace. He looked at me and replied that there is nothing in medicine or science that reaches that level.

To my surprise, he actually asked me if I would speak before the hospital support groups they offered for TBI patients and their families, as he felt my story was very inspirational. He stated there was no way anybody could ever tell I had suffered a brain injury. I enthusiastically replied that I would, but I also assured him that I would not lie if any of them asked how I healed. I would be using the words homeopathy, chiropractic, Reiki, Native American healers, herbs, phosphatidyl choline, etc. My terms were that I would be discussing good nutrition and no processed foods -- organic, if at all possible -- and good, clean water. Also, the use of vitamin and mineral supplements would enter the conversation.

I suggested to this very nice physician that he probably should get approval from his hospital attorneys first, before he had me speak to any group. He assured me his hospital was on the cutting edge and open-minded, and I would hear soon from him. He wrote down my name, address and phone number. He talked to me about the nights and times his support groups were held and when asked if he could refer people to me, I agreed he could. We shook hands, and he walked me to the elevators and out the door. As it turned out, I never heard from him again, and I certainly was not surprised.

This little story says a lot. First of all, it is possible to heal from a (M)TBI without a whole array of drugs. I know because I did it. I was very fortunate. I do not advocate to those who have physicians treating them, discard medications and information that they give you. Absolutely not. But I do suggest to you to "think outside the box" to try other things which may help you also, alternatives or complementaries which can be nontoxic and healing. I suggest that you research the medications you are on and talk with your doctor about them so that you are a partner in your own health care. Many of the things that I used complement conventional medicine and can be good supports.

In addition to focusing on the patient with (M)TBI, never forget that families need education and support -- especially if there are children involved. This is such a difficult thing to live through. My children were worried, frightened and didn't understand what was going on for years. They had lost their mother. But how are families to receive (M)TBI education when their injured family member does not even get diagnosed properly? Physicians really need to spend time with their patients. As discussed earlier, only extensive, irreversible physical brain damage shows up on *some* tests. What about the people suffering from MTBI? They fall through the cracks. Are they not deserving of help? Physicians today rely so heavily on medical tests for diagnostics that the art of sitting with a patient and diagnosing through direct observation and examination seems to have gone the way of the old-time country doc who once did home visits with his black bag. Whatever happened to the art of physical examination and knowing signs and symptoms? Why is this type of diagnostic skill no longer valued, respected, trusted or reimbursed by insurances? There is nothing wrong with all these technological tests, but they are not the "be all and end all," either. The current office visit time allotment of eight to ten minutes per patient for a busy medical office just does not cut it, either.

By now, you may wonder how many people have been injured in a car accident, for example, and have "never been the same since," who are now treated with a symphony of psychiatric drugs or who have taken to using recreational drugs, or alcohol, in order to deal with a post-concussive life. It is frightening to comprehend how many lives, marriages and families are destroyed merely because there are no supports in place.

Above all, show consideration, lots of patience and allow considerable time to anyone who fell, experienced a car accident, returned from a war, has been traumatized, or had a sports injury and "has not been right since." This person needs an advocate and a spokesperson, especially if he/she is silent, or not acting the same and maybe even acts confused. Chances are they might be suffering from aphasia and cannot explain what the problem is to their doctor. My own chiropractor said that his first clue that someone might have a brain injury is that they are very poor historians.

When asked what happened or what hurts, they are vague and just sit there with a blank expression, or might answer with a word or two.

If your loved one is acting this way, go with that person to their physician and speak for him or her. A very changed behavior and personality is what you need to report if this person cannot speak for him or herself. Also, changes in sleeping patterns, irritableness, flying off the handle, crying episodes, rages, an inability to cope with what used to be normal routines, becoming socially isolated, fearfulness, slowness to respond or to answer, confusion, shifting emotions, poor memory, headaches, nausea, dizziness, light-headedness, neck pain, problems with vision, ringing in the ears, loss of taste, or of smell, and an increased sensitivity to lights, sounds or people. The inability to multi-task, difficulty with concentration or paying attention, lack of organizational skills, and the inability to initiate, continue and complete projects needs to be reported. If they do have a brain injury, they will not be able to remember, or to connect to words to tell a doctor these things, or even remember what the physician tells them. If you cannot go with them to their physician appointments, call the office and speak with the nurse or the physician to make sure you know what happened at the office visit and to report these things yourself. In addition, do independent research and educate yourself about (M)TBI and the current modalities used to treat it, as well as alternative ways to help. Advocate for your loved one. If this is a Workers Comp case, get an attorney right away. Screen phone calls to prevent and decrease insurance company harassment that can further harm the injured person. Report this harassment to your attorney, as well as make complaints to the attorney general, your congressman, senator, etc.

What more can you do for someone like this, besides getting them to a doctor? Be kind, give them extra time and be patient. Recovery takes a very long time. If feasible, get help in the house to accomplish simple chores and also to assist the brain-injured person with personal care, and to make sure they are eating, drinking and getting rest. Serve homemade, not processed foods, organic whole foods and clean filtered water. Provide a good, balanced vitamin and mineral supplement. Research phosphatidyl choline and see if that might be useful. Provide and encourage natural foods that are high in this substance.

Make sure they are getting the follow-up medical care they need. Discuss the patient's behaviors you see at home with the doctor. Research and discuss their medications with the doctor. Take notes to refer to later on. Remind them to get up out of bed and take them for short walks if they can tolerate that. Have them sit outside for brief sessions instead of keeping them in the house all the time. These are little things that make all the difference in the world when the person cannot do this independently. Allow and encourage lots of rest time. Be forgiving when their emotions get out of control, but allow the patient his emotions. Encourage them to talk and explain or express what is upsetting them. Decrease the amount of stimulation in their environment, such as lights, music, sounds, confusion, noise and people. Be understanding and supportive if they experience brain fatigue. Try to recognize signs of the patient tiring and encourage rest -- before their emotions explode.

Should you need to discuss something very important with this person, make sure they are rested and turn off unnecessary lights, music, TV, and noise. Don't be busy with their personal care and talking to them at the same time if you need a response. One person should speak at a time.

Stick to the same subject when you talk and allow them a long time to find words to respond. Possibly a communication board can be used if the person cannot speak or write. (This is a simple cardboard tool you can make. Draw or glue pictures of things that the aphasic person can point to: different facial expressions with simple descriptions underneath the pictures [sad, angry, upset, happy, unhappy, worried, in pain, tired, etc], along with pictures of a bathroom, toilet, sink, bed, food, beverages, medicine, etc.) Make lists and write things down for them to reference later. Have them assist you in keeping a journal and calendar. Remind them that these lists and calendars exist, as it is likely they will forget. Have them assist you in doing small things when they are able. Plan and pace their activities, so as not to overtire them.

Daily gentle stretches and exercises as allowed by the physician are also good. Gentle yoga is great for helping to balance the nervous system, as well as T'ai Chi if this is safe for the patient (ask the physician). Allow time for their children to spend with them and allow stress-free family time. Provide emotional support. Use kind words, because your words are "medicine." Make sure you leave them with good thoughts before you physically leave them. I can't emphasize this enough.

If you are within reach of a good structured day program to get them socially involved and re-integrated socially again, that would be great. If that is not possible, provide some structure and routine to their day. Supply them with some short, gentle meditations that you can read aloud to them slowly. Or read poems they may like. Take them through a guided meditation if they are not able to meditate on their own. Get recorded sounds of nature to listen to if you do not live out in the country. When they are able, play a simple game of cards with them, or color or paint with them. Take them for a short walk to look at the flowers or plants growing.

Massage and craniosacral therapy is good. Reiki is wonderful too. If you are interested in homeopathy, consult with the National Center for Homeopathy in Virginia to find a qualified homeopath.

Pray for them and with them if that is something they will agree to. With their permission, contact their priest, minister, clergyman/woman, rabbi, or spiritual advisor. A good psychotherapist can be very helpful to a person with a brain injury as well as their family. This is a tough time for the whole family and supportive services are necessary if you can afford that. Look for the blessings and opportunities that come your way through others entering into your life. Keep a gratitude journal. Keep a daily list of the good things to be thankful for that occur each day. Encourage the injured to express their feelings in ways they are able. Keeping a journal is a great idea if they can write or type. Art work is good too. Music can either be soothing and expressive of emotions or it can be another form of stimulation not tolerated. It all depends on the person and on the day.

"Perseveration" is a common characteristic of a brain-injured person. Basically it is when a person talks a subject to death, repeats and repeats it and can't let go of an idea. According to Merriam-Webster's Online Dictionary: "The continuation of something (as repetition of word) to an exceptional degree or beyond a desired point." This is considered a negative residual symptom of the brain injury. You may notice a person with (M)TBI doing this. Although a physical brain injury is the reason, I would like to offer you my personal view. This physical brain injury also affects memory and so this person may not remember and thus repeats. However, keep in mind that once a person has experienced a lengthy aphasia, and finally can connect to words, he/she wants to communicate and makes sure there is no room for misinterpretation. Believe me when I say that aphasia is a horrific condition. Others cannot understand what you are feeling or experiencing and you cannot speak for yourself. You watch and hear them make assumptions about you and what they think you are and nine times out of 10 it is wrong. Unfortunately, with aphasia you, the patient, cannot tell them differently or correct them. So, when speaking is possible again, an (M)TBI person may expound and repeat the same thing until they are given concrete confirmation that they are truly being heard. They may repeatedly miss the subtle, socially acceptable signs coming from others that their point is taken, or that the subject is what others do not want to hear. Thus, they keep on getting their point across. Perseveration can drive those without brain injuries crazy. To me, it makes all the sense in the world.

What to do about it? Give your undivided attention and truly listen to this person speaking. Repeat back what you've just heard and don't leave out the details. Make it clear that you have truly heard and understood. You will see some of that "perseveration" fall by the wayside. Keep in mind, too, that if these same topics surface again and again, possibly they are not resolved for this person, and need to be. You do need a lot of patience -- maybe as much as the person suffering from (M)TBI.

Above all, don't give up. Remember to be good to yourself, too, as the caregiver. This is a long haul that consumes years. Remind the injured person that, with time and effort, gains will be made and things will get better. Show patience, forgiveness and love. Then get ready to receive this new person, this phoenix that will rise from the ashes.

A Meditation

(A meditation from my Indian Brother)

Sit comfortably, but with your back straight, and hands resting on your knees. Take several deep, slow, even breaths.

Start at the toes. Working your way up, consciously relax every muscle and joint in your body. Don't forget the shoulders, pelvis, abdomen, ribs, face, and head. Try to become totally relaxed.

Then please try this meditation, try to visualize all this happening...

I release all my past, fears, negatives, human relationships, self-image, future and human desires to the light.

I am a light being.

I radiate light from my light center (below your abdomen) throughout my being.

I radiate light from my light center to everyone.

I radiate light from my light center to everything.

I am in a bubble filled with light, only light can come to me and only light can be there.

Thank you God for everyone, everything and for me.

Try to spend 10 – 15 minutes visualizing these things happening, filling up with light, being inside a bubble filled with light, all darkness kept out by the light.

Now place your hands right where the pubis meets the thighs and visualize that area filled with light.

Then place one hand on the lower abdomen (below the navel) and visualize that area filled with light.

Leave that hand in place and place the other hand on the solar plexus. Visualize that area filled with light. Then visualize the light from your lower abdomen joining that light.

Move the lower hand to the heart; visualize the chest cavity filled with light, then everything from the feet to the chest cavity filled with light.

Move the lower hand to the throat, visualize the throat filled with light, and then visualize yourself filled with light from feet to throat.

Move the lower hand to the center of the forehead. Visualize the forehead filled with light. Then the entire body filled with light.

Place one hand over each eye. Maintain the visualization of being filled with light.

This clears the charkas.

Now, finally place both hands on the very top of your head. I'm going to ask you to visualize the Virgin Mary and Christ placing their hands on your head. Visualize them both bathed in brilliant white light. Visualize that brilliant white light completely filling your body from head to toe.

Thank Creator for the healing Christ consciousness that is ours by divine birthright.

Finally, this meditation:

I am eternally grateful for the abundance that is mine. Thank you Creator, for loving me, caring for me and giving me all good things. I love you, think your thoughts and do only what you wish me to do.

Then the Lord's Prayer, broken down (you have never seen it this way):

Our Father who art in heaven and our Mother who art the Earth that nurtures us, hallowed be thy names. Thy kingdom come within me, thy will be already done in Earth our Mother as it is in heaven within us. Give us this day our daily bread as we receive from giving in abundance. And forgive us our errors and misgivings AFTER we have forgiven all others theirs. Lead us not into temptation, but deliver us from the darkness within us. For thine and mine is the kingdom, the power and the glory for my Father in heaven, my Mother the Earth and all living beings are One.

Amen.

A Homeopathic Appendix -- The Consult

It may be difficult to find remedies to help someone with a brain injury, but it is not impossible. Aphasia, memory problems and comprehension difficulties will make this a real challenge. You have to become a very keen observer of body language, and a perceptive listener of what is being said, and what is not being said. You need to listen to the family or the friend who accompanies this client as well. You most likely will hear that this person has never been the same since the injury. You will understand, by observing a family member's tearful or fearful eyes that this is indeed a painful thing to discuss.

Consulting with the family or significant other is a must, too. Ask them what this person was like before the injury. What kind of work did they do? What kind of relationships did they have? What kind of personality did they have? Then, ask them to contrast and compare that former picture with what behaviors, cognition and personality they see now. Make sure to have the family member or their significant other take notes, and enlist their help. He or she, along with the client, are your partners in this case-taking and for future follow-ups. Expect to give more time and patience to these types of clients and cases.

I would suggest you first gather information. Establish that this person is oriented to the here and now. Ask them their name, the year, what the date is, what season is it, who is with them, where they are and why they are here. Ask the history and details of the injury or illness, as well as obtain a complete medical history. Are they on any medications or supplements? Are they using any over-the-counter medicines? For what symptoms are these being used? What is the date or onset of this illness/injury? It is important to know if this brain injury comes from the trauma of an auto accident, or from a war, from a fall, from a sports injury, from a stroke caused by high blood pressure, uncontrolled diabetes, or from dementia, Alzheimer's, or from a disease such as meningitis, etc.

Get as much information as possible. Was there any loss of consciousness or not? Was there a lot of fear at the time? Was there a sense of unrealness? Do they still feel like things are unreal? Do they still experience fear? Ask about physical sensations or injuries that have accompanied this brain insult. Are those sensations and other injuries still present? Were there, or is there still, any numbness, paresthesias or any paresis or paralysis? What about seizures? Find out the client's perceptions, thoughts and emotions at the time. Ask the client what he/she remembers. This fact-finding may be the easiest part of the consult.

After you get all the details of the injury/illness, find out about what is happening now. Is this client being supported by family and friends, or is there little to no support? Is this client caught up in a system such as Workers Comp, or other systems? Is this client being harassed, being treated poorly by others and suspected as a malingerer? What is this client's attitude and response to these tactics and treatment? Ask open-ended questions and let this client talk. What needs to be helped *now*?

Consider everything you see and hear as rubrics for this case. Watch carefully -- how does this person sit and walk? Are they acting fearful or defensive? Do they act overly familiar toward you or anyone else? Are they making any gestures or repetitive gestures? Do they repeat or use the same words? Are all their words appropriate? Are they falling asleep while listening to you? Do they not answer at all but just give you a blank stare and silence? Are they perseverating? Do they have tangential, unfocused answers and thinking? Are they confused as to where they are, how old they are, or might they even be hallucinating or think that they are living in the past? Are their answers too vague? Are there too few words or just silence? Does it appear that this person is absorbed in something you cannot see or comprehend?

Watch and listen carefully and note how your open-ended questions are handled or not handled. Are you getting the information you need or have you stumbled upon a roadblock? Brain-injured persons are noted for being very poor historians. Most have one or all of these challenges: poor memory, altered perception, poor comprehension, and receptive and/or expressive aphasia. You should allow a great amount of time for this person to answer.

You may have to switch gears if this classical, open-ended questioning approach is not working out as well as hoped for. But – remember – this very difficulty is the nature of a brain injury!

Communication Difficulties -- Aphasia

Here are a few suggestions to help your aphasic client communicate with you, if they are not doing well with open-ended questions:

1) If your client can tolerate it and is able to write, ask him/her to keep a journal of his thoughts and feelings for at least a week. This will give you additional baseline information. If this is not appropriate at the beginning, it may prove very useful later on during their recovery, as their brain function improves. (I highly recommend journals as a means to vent and express their feelings – whether they choose to share what they write with anyone or not.) It will be especially useful if your client is having a difficult time with verbalizing thoughts. Ask their significant other to remind the client to journal and write down his/her concerns, and thoughts he would like to share with the homeopath. At the client's home, paper and pen should be kept nearby for the client's use. Short-term memory deficits are a huge problem with someone who is under stress or has a brain injury. Reminders and supplies will have to be supplied by the caregiver.

2) It would be a good idea to have their caregiver document the client's day and responses also. Compare the two journals.

3) It may be best to have two or three short consult sessions to constitute an initial consult. Make sure you have no interruptions, and keep all other sensory stimulation to a minimum. When your client speaks, give your undivided attention, and do not interrupt or they will totally lose their train of thought and concentration. If there is an interruption or other

sensory stimulation, note the effect it has on your client and how he or she responds.

4) Ask some simple, concrete questions that will let you into this person's inner world, and can be answered with a simple "yes" or "no". I wish someone had done that for me when I was suffering from a TBI. How therapeutic and what a relief it is to communicate things that you are experiencing and are worried about, but you have no words, focus, or energy for! It is a great feeling to know you are understood. A brain injury leaves a person feeling so isolated. Unfortunately, not many people know what the inner world of a brain-injured person is like, but I have that advantage. We have to go a few steps further to help this person.

Be hopeful that your client can at least communicate verbally in a "yes" or "no" manner. If not, you may have to use a communication board or point to words or pictures. You can ask them to write their answers, or point to, or circle the words "yes" or "no." If they have difficulty communicating, there are rubrics for aphasia, but it will help you even more to get inside of their world for more specific rubrics.

Simple Questions

Here are some simple questions that you can ask your brain-injured client if they have some form of aphasia and seem unable to answer your open-ended questions. Go slow with these. Be observant of the effect your questions have on your client and let his/her needs guide you with this. Always ask with therapeutic sensitivity and show respect. Some of these questions may be better asked after a few consults, when you have developed a rapport and built trust with your client.

Suggested questions:

Are you having a hard time finding words for how you feel or for what you are thinking?

Are you having a difficult time with your memory?

When you read, can you remember what you have read?

Do you remember what people say to you?

Do you always understand what is being said to you or what you read?

Do lights or sounds bother you, especially if they are happening at the same time? Can you listen attentively to a person speak with lights, music or TV on in the background, or is that more difficult? Is it easier to be with one person talking or with more than one person talking?

If you go into a room, or a place such as a store, can you remember what you are there for? Can you find and pick out the items you need? Are you able to determine if the cashier has given you back the right change when you pay for something?

Are you able to sleep all night?

How do you wake up? Relaxed and rested? Or in a startle, like you have been violently shaken awake?

Do you think differently since your injury/illness?

Do you act differently since your injury/illness?

Can you describe what is different?

Do you dream at all? If yes, what of?

Are you fearful?

Are you angry?

Do you rage?

Are you able to control your emotions?

Do your emotions change quicker now than before the injury?

Do you ever feel embarrassed later on because of your emotional reaction to others?

Do you feel like you have a different personality or that you are different since your injury/illness? (note the client's opinion and perspective of this, as well as their significant other's.)

Are you having a hard time relating to your family and friends since your injury?

Do you prefer to stay alone?

Are you happier with company?

Are you grieving the parts of your self and your life that seem to be gone since this injury?

Do you feel better with consolation or worse with it?

Are you able to cry?

Do you have difficulty finding words to express yourself?

Do you sometimes pick the wrong words?

Do your thoughts seem to spread out all over the place and then disappear?

Can you follow and remember a discussion? Can you follow a story line in a movie or in a book?

Are you having problems with dizziness?

Does your brain ever feel "tired" and/or "bottom out" and you need to be alone? Does this usually happen when you are tired or when light, sounds and people are just too much?

Are you having seizures?

Are you having problems with your balance? When does that happen? Which direction/way do you seem to lean or fall?

Do you ever feel very restless?

Do you ever feel depressed?

Are you ever afraid?

Do you get very sad?

Are there certain times of day or night when these moods are worse?

What makes you feel better?

What makes you feel worse?

Do you laugh a lot? Is your sense of humor different than before your injury/illness?

Are you full of tension or frustration?

Do you feel supported by people in your life?

Do you feel hopeless?

Do you ever wish you were dead?

Do you feel that you cannot communicate your thoughts and feelings as you once did?

What is the best way for you to communicate now?

Do you use bad language more now than before the injury/illness?

Are you more argumentative or irritated now than before?

Are you apathetic?

Is your sex drive different?

Are you more or less demonstrative sexually or sensually now than before your illness/injury?

Do you feel the emotions of jealousy and envy to be stronger, even though it makes little to no sense?

Do you ever feel exhilarated or what you would even call ecstatic at times?

Do you notice feeling more affectionate or more indifferent now than before?

Do you feel you have clairvoyance from this injury/illness?

Do you feel warmer or colder than before your injury/illness?

Do you understand the world in a much different way since your injury/illness?

Does your mind feel either empty of all ideas or too full at times?

Do you have numbness or feel pain anywhere or alternate between the two?

Do you experience headaches or pain in your brain?

Do you feel like you are drifting or floating without a body sometimes?

Do you ever feel like you are anesthetized? Or not connected to this earth anymore? Or that you are floating?

Are you ever confused about how you got an idea -- if you heard it, or dreamt it, or thought it, or if you or someone else said it?

Do you feel lost in a haze or daze?

Are you ever filled with fear?

What are you most fearful of?

Have you ever been afraid that you might harm yourself and not know it?

Have you ever feared that you are going crazy?

Do you forget to eat, or are you disinterested in eating, or do you just have no appetite?

Do you eat more than before your injury/illness?

Do you feel the same emotional connection as before your injury to important people in your life (children, parents, spouse, boyfriend, girlfriend, etc.)?

Are you having difficulties with important relationships in your life since your illness/injury?

Do you feel guilty because of this? Do you feel sad because of this? Do you feel angry because of this? Do you feel indifferent?

Have your spirituality and beliefs changed since your injury/illness?

Do you have a sense of timelessness? Or does time seem to drag?

Before your injury, could you remember a short list of simple things? Are you able to remember the same now?

Are you able to start, continue and finish doing one thing at a time by yourself? Can you do more than one thing at a time?

How is your handwriting? Is it harder for you to write? Is it harder for you to spell? Do you make more mistakes?

Do you get lost in your perceptions or in what you are looking at?

Do you get lost in daydreams?

Have your bowel and bladder habits changed since this injury/illness? How?

Is there any particular time of day or night when you feel worse?

Are any of your symptoms worse with cold or heat or wind or drafts?

Do you notice a difference with a full moon or with certain seasons, or with changing of the seasons?

For women: are your menstrual periods the same as before your illness/ injury? Have they stopped?

Altogether, the information you will gather about the illness/injury itself (the homeopathic etiology), your observations of this client, his/her body language, and the answers you get to these questions should help you with your choice of rubrics. Remind the client and their significant other that they should write things down if they think of something later on after the consult that they want noted.

Always remember that this person who has these odd behaviors, some of which may be considered "anti-social, unfeeling, overly emotional or not appropriate" is not acting this way on purpose. Please do not be judgmental.

Rubrics and Remedies

Here is a list of remedies that are prominent for helping brain injuries. Keep in mind that this list is in no way conclusive or complete. As the great Dr. Hahnemann said: *"Similima similibus curentur."* (Let similars be cured by similars.) Whatever fits the case is what needs to be used.

Aconite	Anacardium
Arnica	Aurum
Belladonna	Baryta carb
Calcarea carb	Cannibus-indicus
Causticum	Conium
Gelsemium	Helleborus
Hyoscyamus	Hypericum
Ignatia	Lachesis
Lycopodium	Natrum sulph
Nitricum acidum	Nux vomica
Opium	Phosphoricum-acidum
Phosphorus	Sepia
Staphysagria	Stramonium
Sulphur	Veratrum album

There is no way one book can tell you what remedy will be needed for any particular case. No weekend workshops in homeopathy can do this and there are no shortcuts to finding a good remedy. Homeopathy is not some "new age" cure-all. This is where many years of experience (some say at least 15 years), and good education come in. A good homeopath is a must. Homeopathy is both an art and a science. The totality of the case must be taken into consideration, as well as the etiology.

If this is an acute situation, consider the homeopathic first-aid remedies: Arnica, Aconite, Belladonna, Natrum Sulphuricum, and Hypericum, etc. It is a well-known fact among homeopaths that it is possible to get good results from treating even old injuries as if they are acutes. Chances are good that you may have a much improved, less muddied symptom picture to work with, after the first-aid remedies have done their work.

Of course, your remedy selection will be dependent upon the symptoms that are being expressed, and the kind of trauma that was sustained.   For example, Hyoscyamus is known to be helpful to those who experience personality changes once they have sustained a brain injury. Homeopathic opium might be helpful for those who seem "stuck" in the traumatic experience and are reliving it, yet also they seem "out there" -- lost in a cloud. Aconite, Argentum nitricum, or Arsencium, are good choices for fear, depending on whether the case fits the remedy. Hypericum is just one example for nerve involvement and injured nerves. If you research the above remedies, you will see that most of them act directly on the brain and nervous system. If the case fits the remedy picture, chances are greater for that remedy to work well.

Keep in mind that rather old language is used in most homeopathic repertories. For example, "apoplexy" means stroke. "Convulsions" is sometimes used for epilepsy or seizures. Utilize your homeopathic dictionary. Keep an updated medical dictionary on hand also. Your client or his family will be coming to you with modern technical medical terms as well. Don't miss out on some great rubrics, which may be overlooked, if you do not know terminology. Hopefully, after reading my personal experience with a TBI and aphasia, and being able to utilize a list of yes/no questions for your aphasic patient, you will be closer to understanding where your client is in his/her interior world, and therefore will have a better grasp of what to look for in a good repertory.

I have used the Mind section of **The Complete Repertory**, by Roger van Zandvoort, for this selection of rubrics. There is such great wealth in the Mind section, which will assist you in describing your client and his mental/cognitive/emotional/behavioral symptoms. Do not neglect to look in the Delusions section also. Repertorize your case and consider also smaller, lesser known remedies. (In addition, I highly suggest the book, **Sensations As If-**, by Herbert A. Roberts, M.D.)

Listed below is a good selection of rubrics you may find appropriate once you have the information from your thorough consult. (Do not neglect to research and read the subrubrics of these broad rubrics listed below. This list is not all-inclusive. Please refer to the **The Complete Repertory** for more.)

(For those still in a coma: MIND; UNCONSCIOUSNESS, coma)

HEAD; CEREBRAL HEMORRHAGE (read all the subrubrics)

HEAD; AILMENTS OF THE CEREBROSPINAL AXIS

MIND; AILMENTS from; injuries, accidents

MIND; AILMENTS from; fright or fear; accident, from sight of

MIND; APHASIA

MIND; CONFUSION of (**Here is a goldmine of a rubric for someone with confusion. Subrubrics include epilepsy, on waking, headaches, location, loses way in well-known streets, from mental exertion, spinal complaints etc. **)

MIND; CONFUSION of mind; dream, as if in

MIND; DAY-DREAMING; tendency for

MIND; DREAM, as if in a

For Memory and Cognitive Deficits, consider:

MIND; MEMORY; confused

MIND; MEMORY; weakness, loss of

MIND; MISTAKES, makes

MIND; FORGETFULNESS

MIND; INSECURITY, mental

MIND; PERSISTS in nothing

MIND; PROSTRATION of mind, mental exhaustion, brain fag

MIND; READING

MIND; RECOGNIZE, DOES NOT

For Emotional Lability, PTSD, Depression, etc also consider:

MIND; ANGER, irascibility; tendency

MIND; ANXIETY

MIND; DESPAIR

MIND; FEAR

MIND; HANDLE things anymore, cannot, overwhelmed by stress; weight, wants to throw off the

MIND; HELPLESSNESS, feeling of

MIND; HIDE, desire to

MIND; HIDE, desire to; fear, from

MIND; HELPLESSNESS, feeling of

MIND; GRIEF

MIND; GRIEF; silent (read all of these subrubrics)

MIND; INDIFFERENCE, apathy

MIND; INDIGNATION

MIND; IRRITABILITY

MIND; JEALOUSY

MIND; RESTLESSNESS, NERVOUSNESS (see subrubrics)

MIND; TWILIGHT agg. mental symptoms (USEFUL FOR SUNDOWN SYNDROME)

MIND; SADNESS; DESPONDENCY

MIND; SECRETIVE

MIND; SELF-CONTROL; loss of

MIND; SENSITIVE, oversensitive

MIND; DELUSIONS is a great rubric and contains a wealth of information there. What you cannot find under Mind – go to the Delusions section. Of course, it is noteworthy to say that the homeopathic meaning for the word "delusion" is not the same as the allopathic meaning.

MIND; COMPANY (**The subrubrics for Mind, Company ameliorates or aggravates are invaluable and contain more subrubrics and references for symptoms of meningitis, diphtheria, epilepsy, wandering, desire for solitude, brain disease, etc**)

For Personality/Behavioral Issues and Changes:

MIND: AFFECTIONATE - for those persons who display affection openly.

MIND; BED contains interesting subrubrics for persons who have some behavior issues related to being in bed.

MIND; CHILDISH behavior (there are some great subrubrics for this relating to stroke and epilepsy)

MIND; CLAIRVOYANCE

MIND; CURSING, swearing; curses, desire to curse

MIND; GESTURES, makes (read the subrubrics for specific gestures)

MIND; IDEAS; deficiency of

MIND; IMPULSIVE

MIND; INJURE; fears to be left alone, lest he should, himself

MIND; INJURE; herself, himself, feels could

MIND; INJURE; satiety of life, must use self-control to prevent shooting himself

MIND; KICKS

MIND; KLEPTOMANIA

MIND; SENSITIVE; OVERSENSITIVITY (see subrubrics for light, noise, external impressions, etc)

For those clients who come out of their injury/illness with much laughter and merriness about everything...

MIND; MIRTH, hilarity, liveliness

MIND; LAUGHING; tendency

MIND; LOQUACITY (please read all the subrubrics for your "perseverating" clients and for those who just talk up a blue streak.)

MIND; MUTTERING

For sexual acting out/promiscuous behavior or talk:

MIND; SHAMELESS

MIND; LASCIVIOUSNESS, lustfulness

MIND; NAKED, wants to be

MIND; NYMPHOMANIA

MIND; SATYRIASIS

MIND; LEWDNESS, obscene

MIND; LEWDNESS, obscene; talk

MIND; LIBERTINISM

It is common with brain injuries for there to be other physical problems, besides the emotional and cognitive deficits. The brain is the master organ of the body and controls *everything* : the ability to walk, move all four extremeties, see, hear, talk, smell, taste, eat, swallow, as well as, reason, think, sleep, dream, the emotions, and it controls vital signs such as pulse, blood pressure, and breathing. Seizures are very common with brain injuries. A person can recover from a brain injury and have no seizures at all, only to experience them for the first time many years later. Underlying conditions of some brain injuries, such as diabetes, hypertension, meningitis, or encephalitis, etc., will have to be considered in the totality of the case. In addition, if this client sustained direct trauma and injury to other body parts, you need to address that, too. The other sections mentioned below contain more rubrics to assist you in tailor-making the case to fit the individual.

Under Generals – Look for the day/night/seasonal elements, etc, alternating states, periodicity, apoplexy, arteriosclerosis, chorea, diabetes, hypertension, immobility of affected parts, injuries, falls, bruises, concussion, painlessness or painfulness of complaints, or paralysis. Repeated Paroxysms (useful for seizures), also look under the rubric "Convulsions" for epilepsy or seizures also. There is a rubric called "Convulsive Movements" for your consideration as well. You will find the rubric "Epileptic Aura" also under the Extremeties Section. The Generals section is a gold mine of information.

There is a separate section in The Complete Repertory, just for Vertigo, which is a very common problem with TBI. It includes auras, cerebral diseases, from concussion, from congestion, tendency to fall, vertigo as if falling, and headaches, which are all very relevant to TBIs.

There are two Head sections. One section features complaints/injuries of the brain, cerebral hemorrhage, etc, and sensations associated with the head. This is a section you definitely want to look at. The other Head section deals exclusively with the type of pain a person may be dealing with. Brain pain and headaches are very common with TBI.

The section of Speech and Voice may be very valuable to you as many brain injured clients do have difficulty with speech. You will observe this when you have your consult.

Swallowing difficulties can be located under the Throat section.

Sleeping difficulties and dreams are found under the Sleep Section. Particular dreams topics are found in the Mind section, under the rubric Dreams.

For those clients with paralysis, hemiplegia, or extremety pain or weakness, please refer to the 2 sections for Extremities – you will find rubrics there to help with those issues.

In closing, it is my sincere wish, that this information will be helpful to homeopaths, and that you will be a blessing to those who come to you for help. JJ

JoAnn M. Jarvis, R.N., D.H.M

JoAnn M. Jarvis, R.N., D.H.M

CPSIA information can be obtained at www.ICGtesting.com
Printed in the USA
BVOW02s2231030815

411674BV00001B/38/P